Applied Satisfiability

Applied Satisfiability

Cryptography, Scheduling, and Coalitional Games

Xiaojuan Liao
Chengdu University of Technology, Chengdu
China

Miyuki Koshimura
Kyushu University, Fukuoka
Japan

WILEY

Published by John Wiley & Sons, Inc., Hoboken, New Jersey.
Published simultaneously in Canada.

Library of Congress Cataloging-in-Publication Data applied for:

Hardback ISBN: 9781394249787

Cover Design: Wiley
Cover Image: © Luvricon/Shutterstock

Set in 9.5/12.5pt STIXTwoText by Straive, Chennai, India

Contents

Preface

Many real-world problems are difficult to solve partially because they present computational challenges. Moreover, it is often important to find not just any solution to the problem, but the optimal one from all feasible solutions according to some objective. In such cases, the problem falls into the class of optimization problems. If an optimization problem is represented as discrete variables, it is known as a combinatorial optimization problem. Boolean satisfiability (SAT) and maximum satisfiability (MaxSAT), as advanced tools for solving combinatorial optimization problems, have been applied to a wide range of practical combinatorial optimization problems.

Historically, there have been a variety of SAT and MaxSAT applications, including bounded model checking, planning, software verification, combinatorial designs, and more. This book extends the application scope of satisfiability theories, demonstrating that high-performance satisfiability-based solvers have great advantages in handling a wider range of problems, provided that these problems are appropriately encoded as propositional formulae. Fundamental questions such as "What new fields can satisfiability theories be applied to?" and "How can advanced satisfiability techniques be effectively utilized to solve practical problems?" are explored.

The application fields introduced in this book primarily encompass encryption key recovery in side-channel attacks, scheduling in single- and parallel-machine environments, and coalition structure generation in coalitional games. These complex problems, challenging to tackle with conventional methods, demonstrate increased efficiency when encoded into Boolean propositional formulae in a proper way. Further details on these applications are presented below.

Satisfiability applied to cryptography: The advanced encryption standard (AES) is a specification for the encryption of electronic data established by the U.S. National Institute of Standards and Technology (NIST) in 2001, and it currently serves as a globally prevalent symmetric cryptographic algorithm. As a kind of block ciphers, AES is vulnerable to a side-channel attack known as a cold boot attack, where the attacker can extract the encryption key from memory. This book exploits the capabilities of SAT and MaxSAT solvers to reconstruct corrupted AES keys extracted from a cold boot attack. This approach is well suited for real attack scenarios where decaying and reverse flipping errors coexist.

Satisfiability applied to scheduling: Scheduling, a recurring decision-making process in various manufacturing and service industries, involves the allocation of resources to tasks over specified time periods. Its primary objective is to optimize one or more goals, such as makespan and total tardiness. This book demonstrates that by formalizing a scheduling problem as a set of Boolean formulae, the problem can be efficiently tackled using the available SAT or MaxSAT solvers in an out-of-the-box manner without further knowledge in scheduling or coding experience from the user. The addressed scheduling problems include job-shop scheduling, overloaded scheduling, and restricted preemptive scheduling. By leveraging satisfiability solving techniques, promising results can be achieved.

Satisfiability applied to coalitional games: Coalition structure generation (CSG) stands as a central research challenge in the coalition formation of multi-agent systems. It involves partitioning a set of agents into exhaustive and disjoint coalitions, where each coalition is assigned a positive or negative payoff. This partition is termed a coalition structure. The objective in solving the CSG problem is to find a coalition structure such that the total value of all the coalitions is maximized. This book shows that by encoding constraints of the CSG problem into Boolean propositional logic and employing an off-the-shelf MaxSAT solver to identify the optimal solution, the solvable problem can be significantly scaled up. In addition to various encoding strategies applied to the CSG problem, this book provides both theoretical and experimental comparisons to explore the underlying reason for performance variation. Building on these insights, a more compact encoding is constructed to further enhance the solving efficiency.

We anticipate that the solving efficiency of the provided encoding methods will see further improvement in the coming years, given the rapid development of novel technologies emerging in SAT and MaxSAT communities. Additionally, this book aims to enlighten researchers and practitioners seeking effective ways to address practical optimization problems. It offers a broader perspective, showing that satisfiability theories can be a competitive alternative for tackling optimization problems. Furthermore, researchers dedicated to the application areas covered in this book may find occasional utility in its insights.

December, 2024

Xiaojuan Liao
Chengdu, China

Miyuki Koshimura
Fukuoka, Japan

1

Satisfiability Theories

After reading this chapter, you should be able to:

- Comprehend fundamental concepts of Boolean satisfiability (SAT), maximum satisfiability (MaxSAT), and MaxSAT extensions.
- Understand mainstream algorithms for solving SAT and MaxSAT problems.

1.1 Boolean Satisfiability (SAT)

A Boolean formula is a string that represents a Boolean function. A Boolean function is a function of the form: $B^k \to B$, where $B = \{True, False\}$ is a Boolean domain and k is the arity of the function. Usually, *True* and *False* are represented by 1 and 0, respectively. A propositional Boolean formula is a Boolean formula that only contains logic operations *and*, *or* and *not* (sometimes called *negation* or *complement*), symbolized as $\wedge$, $\vee$, and $\neg$, respectively. Some examples of propositional Boolean formulas are listed as follows:

$$(a \wedge b) \vee (a \wedge \neg c) \vee (b \wedge c \wedge d)$$
$$(a \vee b) \wedge (a \vee \neg b \vee \neg c) \wedge (c \vee \neg a)$$
$$(b \vee \neg c) \wedge (a \vee \neg c) \vee b$$

A propositional Boolean formula can be expressed in conjunctive normal form (CNF), also known as *product of sum (POS)* form. A formula in CNF consists of a conjunction (logic *and*) of one or more clauses. A clause is a disjunction (logic *or*) of one or more literals, and a literal is an occurrence of a Boolean variable or its negation. An example of a propositional Boolean formula in CNF is $\phi = (b \vee \neg c) \wedge (a \vee \neg c) \wedge b$, which can also be represented by a set of clauses as $\phi = \{b \vee \neg c, a \vee \neg c, b\}$, or an assemble of literal sets like $\phi = \{\{b, \neg c\}, \{a, \neg c\}, \{b\}\}$. Usually, a CNF formula is denoted by a set of clauses, i.e., conjunction is omitted.

Applied Satisfiability: Cryptography, Scheduling, and Coalitional Games, First Edition. Xiaojuan Liao and Miyuki Koshimura.

The problem of determining whether there exists a variable assignment that makes a propositional Boolean formula evaluate to true is called a Boolean satisfiability (SAT) problem. In other words, the SAT problem tries to find a variable assignment to a CNF formula that satisfies all the clauses in a Boolean propositional formula. If such an assignment exists, the formula is *satisfiable*; otherwise, the formula is *unsatisfiable*.

Example 1.1 $\phi = \{a \vee b \vee \neg c \vee d, a \vee \neg b \vee \neg c, a \vee b, \neg d, d\}$ is a Boolean propositional formula in CNF. This formula contains five clauses: $(a \vee b \vee \neg c \vee d)$, $(a \vee \neg b \vee \neg c)$, $(a \vee b)$, $(\neg d)$, and (d). The first clause $(a \vee b \vee \neg c \vee d)$ contains four literals, i.e., a, b, $\neg c$, and d. This formula is unsatisfiable because the last two clauses are conflicting, which means they cannot be satisfied at the same time.

SAT problem is the first known nondeterministic polynomial time (NP)-complete problem proven by Cook [1]. The proof shows how every decision problem in the complexity class NP can be reduced to the SAT problem for CNF formulas. That the SAT problem is NP-complete problem briefly means that there is no known algorithm that efficiently solves all instances of SAT, and it is generally believed that no such algorithm can exist. A class of algorithms to efficiently solve a large enough subset of SAT instances is called SAT solver. Extending the capabilities of SAT solving algorithms is an ongoing area of progress. However, no current such methods can efficiently solve all SAT instances.

In the last century, showing that a certain problem is as hard as SAT was the end of the story and trying to solve it directly seemed to be hopeless. With the significant progress made in SAT solving, it is now widely accepted that being able to encode a problem into SAT is highly likely to lead to a practical solution. This "SAT Revolution" started at the end of the last century and continues to produce amazing new practical and theoretical results [2].

Examples of SAT applications include software verification [3, 4], bounded model checking [5, 6], artificial intelligence (AI) planning [7, 8], cryptographic attacks [9–13], scheduling [14–17], etc. These applications rely on the ability of SAT solvers to determine whether there exists an assignment that makes a given Boolean formula evaluate to true and return one satisfying assignment if the formula is satisfiable. It is noticeable that there are applications that require enumerating all the satisfying assignments. For example, in frequent itemset mining, it is necessary to generate all sets of items with high support from a given transaction database [18]. This opens up a new variant of SAT, called all solutions SAT (AllSAT). Solving an AllSAT problem aims to enumerate all satisfying assignments if a given CNF formula is satisfiable. Readers may refer to [19] for major techniques of ALLSAT solvers.

1.2 Maximum Satisfiability (MaxSAT)

Given a Boolean propositional formula, if it is unsatisfiable, SAT solvers only report that no solution exists, without any information on unsatisfiable instances. However, assignments violating a minimum number of constraints, or satisfying all the compulsory (*hard*) constraints and as many optional (*soft*) constraints as possible, can be considered as acceptable solutions in real-life scenarios. To cope with this limitation of SAT, maximum satisfiability (MaxSAT) and its extensions, such as partial MaxSAT and weighted MaxSAT, are becoming an alternative for representing and efficiently solving over-constrained problems [2].

The MaxSAT problem for a CNF formula is the problem of finding a variable assignment that maximizes the number of satisfied clauses. MaxSAT is often used to mean MinUNSAT because finding an assignment that maximizes the number of satisfied clauses is equivalent to finding an assignment that minimizes the number of unsatisfied clauses. MaxSAT is useful to measure the extent of unsatisfiability of a CNF formula.

Three extensions of MaxSAT are more well suited for representing and solving over-constrained problems: partial MaxSAT, weighted MaxSAT, and weighted partial MaxSAT (WPM).

In a partial MaxSAT instance, each clause is labeled either *hard* or *soft*. The hard clauses must be obligatorily satisfied, while the soft clauses can be unsatisfied. The goal of solving the partial MaxSAT instance is to satisfy all hard clauses and the maximal number of soft clause. The partial MaxSAT problem is easily extended to a SAT problem if all the clauses are hard, and a MaxSAT problem if all the clauses are soft.

Example 1.2 Given a partial MaxSAT instance $\phi = \{[a], [\neg a \vee \neg b], (b \vee c)\}$, the first and second clause, enclosed by "[ ]," are hard clauses, and the third clause, enclosed by "( )," is a soft clause. To satisfy the hard clauses, a and b are forced to be 1 and 0, respectively. Based on the assignment of a and b, the true assignment of c has to be 1 so that the soft clause can be satisfied.

A weighted MaxSAT instance is expressed in weighted CNF, where each clause is assigned a positive integer. The problem is to find a truth assignment that maximizes the sum of weights of satisfied clauses. In a special case, if the weights of all the clauses in weighted MaxSAT are equal to one, the problem is regarded as a MaxSAT problem.

Example 1.3 A weighted MaxSAT instance $\phi = \{(a, 2), (\neg a \vee \neg b, 3), (b \vee c, 4)\}$ has three weighed clauses holding weights 2, 3, 4, respectively. All of the three clauses can be satisfied by assigning 1, 0, 1 to a, b, c, respectively, which leads to the maximal sum of weights of satisfied clauses.

The WPM is the combination of partial MaxSAT and WPM. A WPM instance distinguishes hard and soft clauses, where each soft clause is assigned a positive integer. Solving the WPM instance is to satisfy all hard clauses and maximize the sum of weights of satisfied soft clauses. The definition of WPM can be easily extended to partial MaxSAT, where all soft clauses have weight 1, and weighted MaxSAT, where no clauses are hard.

Example 1.4 Given a WPM instance $\phi = \{[a], [\neg a \vee \neg b], (b \vee c, 2), (b \vee \neg c, 7)\}$, the first and second clause, enclosed by "[ ]," are hard clauses, while the third and fourth clauses, enclosed by "( )," are soft clauses. To satisfy the hard clauses, a and b are forced to be 1 and 0, respectively. Based on the assignment of a and b, the true assignment of c is preferred to be 0 so that the weight 7 can be earned, which is larger than the other case that the gain is merely 2.

Many important problems can be naturally expressed as MaxSAT, including academic problems such as Max-Cut or Max-Clique, as well as problems from many industrial domains. Concrete examples include the following domains: routing problems [20], hardware debugging [21–23], software debugging [24, 25], scheduling [17, 26–28], planning [29–32], coalitional games [33–37], etc. Additionally, many problems originally formulated in other optimization frameworks can be easily reformulated as MaxSAT, such as the Pseudo-Boolean Optimization framework [38], the Weighted Constraint Satisfaction Problem (WCSP) framework [39], and the MaxSMT framework [40]. Readers may refer [41] for more traditional applications of MaxSAT.

1.3 Satisfiability Algorithms

The last two decades have witnessed significant progress in the development of theoretical, logical and algorithmic aspects of SAT and MaxSAT solving techniques. Moreover, the SAT competition[1] (starting from 2002), MaxSAT evaluation[2] (starting from 2006), and the international conference on theory and applications of satisfiability testing[3] (first held in 1996) jointly play as a driving force for motivating the development of novel SAT and MaxSAT technologies.

1 http://satcompetition.org.
2 https://maxsat-evaluations.github.io.
3 www.satisfiability.org.

1.3.1 SAT Algorithms

Algorithms for solving the SAT problem are mainly based on incomplete algorithms and complete algorithms. Incomplete algorithms (e.g., GSAT [42], WalkSAT [43]) can get a satisfying assignment quickly, but do not prove unsatisfiability if the formula is unsatisfiable. By contrast, complete algorithms (e.g., DPLL algorithm and its variants Chaff [44] and Generic seaRch Algorithm for the Satisfiability Problem, GRASP [45]) guarantee to find satisfying assignment or prove unsatisfiability.

Incomplete algorithms are usually based on stochastic local search, which iteratively improve an assignment of the variables until all constraints are satisfied. Stochastic local search algorithms run efficiently especially on large-scale instances. The primary challenge is the cycling problem where a candidate solution may be revisited [46]. Random walk, restarting strategies, and the configuration check [46] are usually used to tackle this problem. For an in-depth understanding of most recent incomplete algorithms, we recommend readers to consult a comprehensive survey on intelligent optimization algorithms [47].

Complete algorithms are based on the modern improvements or variants of the DPLL approach. DPLL [48, 49], named after the four collective authors, i.e., Davis, Putnam, Logemann, and Loveland, is a complete, backtracking-based algorithmic framework that is the basis of many of the most successful modern complete SAT solvers.

The backtracking algorithm takes a CNF formula ϕ as input. Its objective is to determine whether ϕ is satisfiable, and if so to find a satisfying assignment. The algorithm maintains a truth assignment and runs recursively. Initially, the assignment is empty with all variables unassigned. At each recursive step, the algorithm selects a literal l and explores two potential assignments: true and false. This is done by making recursive calls to investigate both assignments to l. The algorithm continues this process until it either finds a satisfying assignment or it exhaustively search through the entire space and concludes that ϕ is unsatisfiable [50].

In the worst case, the backtracking algorithm has to explore the entire search space by setting each literal first true and then false. To enhance the search efficiency, based on the backtracking algorithm, DPLL is enhanced by *unit propagation* and *pure-literal elimination* rules.

If a clause contains only a single unassigned literal, it is called a unit clause. Given a CNF formula, the unit-clause rule checks if the formula contains a unit clause. If so, it removes all clauses that contain the literal from the formula and removes the negation of the literal from the other clauses. Unit propagation is the iterated application of the unit-clause rule, which repeatedly applies the unit-clause rule until either it derives an empty clause or there are no more unit clauses left.

Example 1.5 Consider the following propositional formula in CNF:

$$((a \vee b \vee c \vee \neg d) \wedge (\neg a \vee c) \wedge (\neg c \vee d) \wedge (a))$$

In this formula, (a) is a unit clause and must be assigned true, then it can be removed from the formula. The first clause $(a \vee b \vee c \vee \neg d)$ contains a, thus this clause is always satisfied and can be removed from the formula safely.

Consider the second clause $(\neg a \vee c)$, which contains $\neg a$. Now that we have known that $\neg a$ is false, we can remove $\neg a$ from the clause since it has nothing to do with the truth value of the clause.

After unit propagation, the formula can be simplified as

$$((c) \wedge (\neg c \vee d))$$

Here, we find the simplified formula contains a new unit clause (c). Continue with the unit propagation. We can remove c from the formula and remove $\neg c$ from the clause $(\neg c \vee d)$. Thereby, the formula is simplified as

$$(d)$$

Apparently, (d) is a unit clause and can be removed from the formula. Eventually, the formula becomes empty, and the original formula is satisfiable.

If a literal appears in some clause but its negation does not appear in the formula, it is called a pure literal. Pure-literal elimination removes all clauses containing a pure literal because these clauses can be always satisfied by making the pure literal true without falsifying any other clauses.

Example 1.6 Consider the following propositional formula in CNF:

$$((\neg a \vee b) \wedge (a \vee c \vee \neg b) \wedge (c \vee \neg d) \wedge (d \vee \neg e))$$

In the formula, c and $\neg e$ are pure literals. We can always satisfy the clauses $(a \vee c \vee \neg b)$, $(c \vee \neg d)$, and $(d \vee \neg e)$ by assigning c to true and e to false. After the pure-literal elimination, the formula can be simplified as

$$(\neg a \vee b)$$

This simplified formula contains pure literals $\neg a$ and b. Therefore, we can remove this clause from the formula. Eventually, the formula is empty, and the original formula is satisfiable.

The DPLL procedure is described in Algorithm 1.1. Functions *UnitPropagate*(ϕ) and *PureLiteralEliminate*(ϕ) take a CNF formula ϕ as input and return the simplified formula of applying unit propagation and pure-literal elimination, respectively. By picking an unassigned variable from ϕ, which is accomplished by *SelcetVariable*(ϕ), and assigning truth values to it, the resulting formula is simplified iteratively. This process continues until ϕ is empty or ϕ contains an empty clause, indicating that ϕ is satisfiable or unsatisfiable, respectively.

Algorithm 1.1 DPLL recursive procedure

Input: A CNF formula ϕ.
Output: SAT if ϕ is satisfiable; UNSAT, otherwise.

function DPLL(ϕ)
  while ϕ contains a unit clause **do**
    $\phi \leftarrow$ *UnitPropagate*(ϕ) ▷ unit propagation
  end while
  while ϕ contains a pure literal **do**
    $\phi \leftarrow$ *PureLiteralEliminate*(ϕ) ▷ pure-literal elimination
  end while
  if ϕ is empty **then**
    return SAT ▷ ϕ is satisfiable.
  end if
  if ϕ contains an empty clause **then**
    return UNSAT ▷ ϕ is unsatisfiable.
  end if
  $x \leftarrow$ *SelectVariable*(ϕ) ▷ Select an unassigned literal x
  return $DPLL(\phi_{x \leftarrow true}) \vee DPLL(\phi_{x \leftarrow false})$
end function

The DPLL algorithm has two primary limitations. First, upon encountering a conflict, where a variable assignment renders a clause unsatisfiable, the algorithm identifies only the present variable assignment as the source of the conflict, lacking the ability to deduce specific combinations of variable assignments that may also trigger the conflict. Second, when a conflict occurs, the algorithm employs the chronological backtracking approach, jumping back to the most recent variable assignment that led to the conflict and assigning the opposite truth value to that variable. This strategy may result in a considerable amount of time, as it wastes time exploring variable assignments in a search space that is guaranteed to lead to conflicts.

Conflict-driven clause learning (CDCL) [44, 45], improved from the DPLL algorithm, incorporates conflict analysis and clause learning techniques to avoid repeatedly exploring the same impossible assignments to literals, thus pruning the search space effectively and expediting the solving process. Main improvements of the CDCL are *clause learning from conflicts* and *nonchronological backtracking*, which are summarized as follows:

- Clause learning from conflicts: When detecting a conflict, i.e., a clause becomes unsatisfiable, CDCL analyzes the conflict to identify the cause and adds a learnt clause (the negation of the assignments that led to the conflict) to the original formula. This learnt clause is essential for avoiding the same conflict in the future, as it blocks the problematic assignment.
- Nonchronological backtracking: If a conflict arises, CDCL jumps back to the appropriate decision level where the first-assigned variable involved in the conflict and assigns the opposite truth value to it. The backtracking level may not necessarily be the level where the most recently assigned variable locates. This allows the algorithm to flexibly select the decision level for backtracking and intelligently avoid conflict repetition by skipping over potentially unnecessary variable assignments.

Given the ability to discover early that branches of the search space does not have to be explored, the CDCL algorithm is extremely efficient for solving large-scale SAT problems. Currently, the CDCL algorithm stands out as one of the most efficient methods for solving SAT problems, laying the foundation of many modern SAT solvers (e.g., MiniSAT [51], Glucose [52], MergeSat [53], and IsaSAT [54]). For more details on the CDCL procedure, we recommend readers to refer to [50]. For a comprehensive study and analysis of the latest developments and new approaches of SAT solvers, we suggest referring to [55].

1.3.2 MaxSAT Algorithms

Generally speaking, there are two approaches for MaxSAT solving techniques: heuristic and approximation algorithms that find near-optimal solutions and exact algorithms that compute optimal solutions.

Heuristic local search algorithms are the foundation of early practical works to find near-optimal solutions. Whereas many exact MaxSAT solvers use local search algorithms to rapidly compute an initial assignment of variables, these algorithms do not guarantee the quality of their output solutions. By contrast, approximation algorithms output solutions not as fast as heuristic algorithms, but they provide a guarantee about the quality of their solutions. The approximation of a MaxSAT solution is usually measured by a factor that is bounded by a constant α or a slowly growing function of the input size. Given a constant α, an algorithm is α-approximation for a maximization problem if it provides a feasible solution in polynomial time, which is at least α times the optimum, considering all the possible instances of the problem [56]. A number of improvements have been achieved on the performance guarantee, from 1/2, proposed in 1974 [57], to 0.7584, proposed in 1995 [58]. Later on, a limit on approximability was proved by Håstad [59] that unless NP=P, no approximation algorithm for MaxSAT can achieve a performance guarantee better than 7/8. This theory was proved again in [60], showing that the constant 7/8 is tight. From a theoretical and practical point of view, semidefinite programming has been shown quite promising for approximating MaxSAT solutions. Readers may refer to [61] to learn more about how to approximate MaxSAT with semidefinite programming.

Exact algorithms can be classified into two approaches. The one follows a branch and bound (BB) algorithm and applies several techniques tailored to MaxSAT. Another one makes use of a state-of-the-art SAT solver as an inference engine, referred to as a SAT-based approach.

Many contemporary exact MaxSAT solvers follow a BB algorithm [62–69], which ensures the minimal number of unsatisfied clauses in a MaxSAT problem. Given a MaxSAT instance ϕ, BB explores a search tree that represents the space of all possible assignments for ϕ in a depth-first manner. At every node, BB compares the upper bound (*UB*) with the lower bound (*LB*). *UB* is the best solution (i.e., the minimum number of falsified clauses) found so far for a complete assignment, and *LB* is the sum of the number of clauses which are falsified by the current partial assignment plus an underestimation of the number of clauses that will become unsatisfied if the current partial assignment is completed. If $LB \geq UB$, the algorithm prunes the subtree below the current node and backtracks chronologically to a higher level in the search tree. If $LB < UB$, the algorithm tries to find a better solution by extending the current partial assignment by assigning one more variable. The value of *UB* after the search of entire tree is the optimal number of unsatisfied clauses in ϕ.

BB algorithms usually perform effective in solving small-scale MaxSAT problem instances with a few hundred variables. However, they are ineffective on instances with more than a thousand variables [70]. Currently, the most important algorithms for solving MaxSAT problems are based on iterative calls to a SAT solver, commonly known as the SAT-based approaches. These approaches can be further classified into three types: the model-improving, the core-guided, and the implicit hitting set approaches.

Model-improving approaches (e.g., SAT4j [71], QMaxSAT [72], Pacose [73], and MergeSat [74]) are based on querying a SAT solver for solutions of increasing quality. Given a MaxSAT instance ϕ with a collection of n soft clauses $\{C_1, \dots, C_n\}$, a new variable b_i is added to each soft clause C_i $(1 \leq i \leq n)$, where b_i is called a *blocking variable*. Solving the MaxSAT problem for ϕ is to minimize the number of blocking variables that evaluate to true, called *true blocking variables*, in ϕ' with soft clauses $\{C_1 \vee b_i, \dots, C_n \vee b_n\}$. The minimal satisfied assignment is searched by iterative calls to a SAT solver, which are summarized as follows [72]. First run the SAT solver on ϕ' without any constraints to get an initial model and count the number k of true blocking variables in the model, then add a constraint to limit the number of true blocking variables to a smaller k (called a cardinality constraint), and run the solver again. If the problem is unsatisfied, k is the optimal solution. Otherwise, the process is repeated with the constraint that limits the number of true blocking variables to a smaller integer. This process terminates when the problem becomes unsatisfied.

Model-improving solvers start by adding a blocking variables to each soft clause and then iteratively call a SAT solver by decreasing the number of true blocking variables until the formula becomes unsatisfiable. When a MaxSAT problem has a significantly large number of soft clauses, solving this problem becomes infeasible due to the resulting extensive encodings.

In contrast to model improving approaches, core-guided approaches (e.g., Fu–Malik algorithm [75], WPM1 [76], WMSU1 [77], OLL [78], and Maxino [79]) are unsatisfiability based, working from unsatisfiable (UNSAT) to SAT. Given a MaxSAT instance ϕ, the following process is iterated until ϕ is satisfiable: First run a SAT solver on ϕ. If ϕ is unsatisfiable, extract an unsatisfiable subset $US = \{C_1, \dots, C_m\}$ from ϕ and introduce m new blocking variables b_i $(1 \le i \le m)$. The unsatisfiable subset of clauses is called a *core*. Then, replace C_i with $C_i \vee b_i$ $(1 \le i \le m)$ and add a constraint $\sum_{i=1}^{m} b_i = 1$ to build a new formula ϕ'. The process is called *relaxation*. If ϕ' becomes satisfiable, the iteration terminates. The number of iterations indicates the number of falsified clauses in the original ϕ.

Core-based algorithms have proved to be effective on industrial problems, typically suitable for solving instances that have optimal solutions falsifying very few soft clauses (relative to the total number of soft clauses). A drawback of core-based algorithms is that the relaxation requires adding new cardinality constraints to the formula at each step, which makes the problem harder to solve at each iteration. Narodytska and Bacchus [80] proposed an alternative approach for solving WPM problems, which also builds a sequence of new SAT formulas, but avoids using cardinality constraints. The new core-guided solver outperforms the contemporary solvers by a large margin in the total number of industrial problem instances.

Similar to the purely SAT-based core-guided approaches, implicit hitting set solvers (e.g., [81–84]) extract cores in an iterative fashion. Given a set of cores, a hitting set is a set of soft clauses that includes at least one soft clause from each core. A hitting set is optimal (of minimum cost) if and only if the sum of the weights of the soft clauses in it is smallest among all hitting sets of the set of cores. In contrast to core-based algorithms that transform the input instance using core compilation steps, the input MaxSAT instance of implicit hitting set solvers is not altered during search. Each SAT solver call is made on the original hard clauses together with a subset of the original soft clauses. Thus, the MaxSAT instance does not get larger in size as the search progresses. As a result, the cores found during search remain relatively small compared to the core-guided approaches [70].

Although SAT and MaxSAT problems have a long history, the solving techniques have evolved and improved over the years. Many advanced techniques have been introduced to aid in problem solving, such as parallel computing [74, 85–88], reinforcement learning [88, 89], machine learning [90–92], and quantum computing [93–95]. These new techniques, in turn, encourage SAT and MaxSAT applications in more real-world scenarios.

1.4 Chapter Summary

This section introduced the fundamental formal concepts of SAT, MaxSAT, partial MaxSAT, weighted MaxSAT, and WPM. Mainstream algorithms for solving SAT and MaxSAT were summarized.

From a practical view, many existing applications treat SAT and MaxSAT solvers as black boxes. In other words, once a real-world problem is translated into a CNF formula, the problem can be solved by utilizing an off-the-shelf solver without any interactions between the solver and the application. However, understanding the satisfiability algorithms is beneficial for selecting appropriate solvers, analyzing performance, and further optimizing the efficiency of solving specific problems.

References

1 S. A. Cook. The complexity of theorem-proving procedures. In *Proceedings of the Annual ACM Symposium on Theory of Computing*, pages 151–158, 1971.

2 A. Biere, M. Heule, H. van Maaren, and T. Walsh. *Handbook of Satisfiability* - Second Edition, IOS Press, 2021.

3 E. M. Clarke, D. Kroening, and F. Lerda. A tool for checking ANSI-C programs. In *Proceedings of Tools and Algorithms for the Construction and Analysis of Systems*, pages 168–176, 2004.

4 D. Jackson, I. Schechter, and I. Shlyakhter. Alcoa: the alloy constraint analyzer. In *Proceedings of International Conference on Software Engineering*, pages 730–733, 2000.

5 A. Biere, A. Cimatti, E. Clarke, and Y. Zhu. Symbolic model checking without BDDs. In *Proceedings of Tools and Algorithms for the Construction and Analysis of Systems*, pages 193–207, 1999.

6 M. Sheeran, S. Singh, and G. Stalmarck. Checking safety properties using induction and a SAT solver. In *Proceedings of the 3rd International Conference on Formal Methods in Computer-Aided Design*, pages 108–125, 2000.

7 J. Rintanen, K. Heljanko, and I. Niemelä. Planning as satisfiability: Parallel plans and algorithms for plan search. *Artificial Intelligence*, 170(12-13):1031–1080, 2006.

8 B. Selman and H. Kautz. Planning as satisfiability. In *Proceedings of European Conference on Artificial Intelligence*, pages 359–363, 1992.

9 A. A. Kamal. Applications of SAT solvers to AES key recovery from decayed key schedule images. In *Proceedings of the 4th International Conference on Emerging Security Information Systems and Technologies*, pages 216–220, 2010.

10 C. Patsakis. RSA private key reconstruction from random bits using SAT solvers. *IACR Cryptology ePrint Archive*, 2013.

11 X. Liao, H. Zhang, M. Koshimura, H. Fujita, and R. Hasegawa. Using MaxSAT to correct errors in AES key schedule images. In *25th IEEE International Conference on Tools with Artificial Intelligence, ICTAI 2013*, Herndon, VA, USA, November 4–6, 2013, pages 284–291. IEEE Computer Society, 2013.

12 X. Liao, H. Zhang, and M. Koshimura. Reconstructing AES key schedule images with SAT and MaxSAT. *IEICE Transactions on Information and Systems*, 99-D(1):141–150, 2016.

13 I. Zimerman, E. Nachmani, and L. Wolf. Recovering AES keys with a deep cold boot attack. In M. Meila and T. Zhang, editors, *Proceedings of the 38th International Conference on Machine Learning, ICML 2021*, 18–24 July 2021, Virtual Event, volume 139 of *Proceedings of Machine Learning Research*, pages 12955–12966. PMLR, 2021.

14 J. M. Crawford and A. B. Baker. Experimental results on the application of satisfiability algorithms to scheduling problems. In *12th AAAI National Conference on Artificial Intelligence*, pages 1092–1097, 445 Burgess Drive Menlo Park, CA, United States, Oct. 1994. American Association for Artificial Intelligence.

15 M. Koshimura, H. Nabeshima, H. Fujita, and R. Hasegawa. Solving open job-shop scheduling problems by SAT encoding. *IEICE Transactions on Information and Systems*, E93-D(8):2316–2318, 2010.

16 R. Nieuwenhuis, A. Oliveras, E. Rodríguez-Carbonell, and E. Rollon. Employee scheduling with SAT-based pseudo-Boolean constraint solving. *IEEE Access*, 9:142095–142104, 2021.

17 X. Liao, H. Zhang, M. Koshimura, R. Huang, and F. Li. Solving restricted preemptive scheduling on parallel machines with SAT and PMS. *Journal of Universal Computer Science*, 29(8):911–937, 2023.

18 I. O. Dlala, S. Jabbour, L. Saïs, and B. B. Yaghlane. A comparative study of SAT-based itemsets mining. In M. Bramer and M. Petridis, editors, *Research and Development in Intelligent Systems XXXIII - Incorporating Applications and Innovations in Intelligent Systems XXIV. Proceedings of AI-2016, The 36th SGAI International Conference on Innovative Techniques and Applications of Artificial Intelligence*, Cambridge, UK, December 13–15, 2016, pages 37–52. Springer, 2016.

19 T. Toda and T. Soh. Implementing efficient all solutions SAT solvers. *ACM Journal of Experimental Algorithmics*, 21(1):1.12:1–1.12:44, 2016.

20 H. Xu, R. Rutenbar, and K. Sakallah. Sub-SAT: A formulation for relaxed Boolean satisfiability with applications in routing. In *Proceedings of International Symposium on Physical Design*, pages 182–187, 2002.

21 S. Safarpour, H. Mangassarian, A. Veneris, M. H. Liffiton, and K. A. Sakallah. Improved design debugging using maximum satisfiability. In *Proceedings of the 7th Conference on Formal Methods in Computer-Aided Design*, pages 13–19, 2007.

22 Y. Chen, S. Safarpour, A. Veneris, and J. Marques-Silva. Spatial and temporal design debug using partial MaxSAT. In *Proceedings of ACM Great Lakes Symposium on VLSI*, pages 345–350, 2009.

23 H. Mangassarian, A. G. Veneris, S. Safarpour, F. N. Najm, and M. S. Abadir. Maximum circuit activity estimation using pseudo-Boolean satisfiability. In *Proceedings of Conference on Design, Automation and Test in Europe*, pages 1538–1543, 2007.

24 M. Jose and R. Majumdar. Bug-assist: Assisting fault localization in ANSI-C programs. In *Proceedings of International Conference on Computer Aided verification*, pages 504–509, 2011.

25 M. Jose and R. Majumdar. Cause clue clauses: Error localization using maximum satisfiability. In *Proceedings of ACM SIGPLAN Conference on Programming Language Design and Implementation*, pages 437–446, 2011.

26 E. Demirović, N. Musliu, and F. Winter. Modeling and solving staff scheduling with partial weighted MaxSAT. *Annals of Operations Research*, 275(1):79–99, 2019.

27 X. Liao, H. Zhang, M. Koshimura, R. Huang, W. Yu, and F. Li. Modeling and solving scheduling in overloaded situations with weighted partial MaxSAT. *Mathematical Problems in Engineering*, 2021:1–17, 2021.

28 A. Lemos, P. T. Monteiro, and I. Lynce. Introducing *UniCorT*: An iterative university course timetabling tool with MaxSAT. *Journal of Scheduling*, 25(4):371–390, 2022.

29 M. C. Cooper, S. Cussat-Blanc, M. de Roquemaurel, and P. Régnier. Soft arc consistency applied to optimal planning. In *Proceedings of International Conference on Principles and Practice of Constraint Programming*, pages 680–684, 2006.

30 F. Juma, E. I. Hsu, and S. A. McIlraith. Preference-based planning via MaxSAT. In *Proceedings of Canadian Conference on AI*, pages 109–120, 2012.

31 N. Robinson, C. Gretton, D. N. Pham, and A. Sattar. Partial weighted MaxSAT for optimal planning. In *Proceedings of Pacific Rim International Conference on Artificial Intelligence*, pages 231–243, 2010.

32 L. Zhang and F. Bacchus. MaxSAT heuristics for cost optimal planning. In *Proceedings of the 26th AAAI Conference on Artificial Intelligence*, pages 1846–1852, 2012.

33 X. Liao, M. Koshimura, H. Fujita, and R. Hasegawa. Solving the coalition structure generation problem with MaxSAT. In *IEEE 24th International Conference on Tools with Artificial Intelligence, ICTAI 2012*, Athens, Greece, November 7–9, 2012, pages 910–915. IEEE Computer Society, 2012.

34 X. Liao, M. Koshimura, H. Fujita, and R. Hasegawa. MaxSAT encoding for MC-net-based coalition structure generation problem with externalities. *IEICE Transactions on Information and Systems*, 97-D(7):1781–1789, 2014.

35 X. Liao, M. Koshimura, H. Fujita, and R. Hasegawa. Extending MaxSAT to solve the coalition structure generation problem with externalities based on agent relations. *IEICE Transactions on Information and Systems*, 97-D(7):1812–1821, 2014.

36 X. Liao and M. Koshimura. A comparative analysis and improvement of MaxSAT encodings for coalition structure generation under MC-nets. *Journal of Logic and Computation*, 29(6):913–931, 2019.

37 X. Liao, M. Koshimura, K. Nomoto, S. Ueda, Y. Sakurai, and M. Yokoo. Improved WPM encoding for coalition structure generation under MC-nets. *Constraints*, 24(1):25–55, 2019.

38 F. A. Aloul, A. Ramani, I. L. Markov, and K. A. Sakallah. Generic ILP versus specialized 0-1 ILP: An update. In *Proceedings of International Conference on Computer-Aided Design*, pages 450–457, 2002.

39 J. Larrosa and T. Schiex. Solving weighted CSP by maintaining arc consistency. *Artificial Inteligence*, 159(1-2):1–26, 2004.

40 R. Nieuwenhuis and A. Oliveras. On SAT modulo theories and optimization problems. In *Proceedings of International Conference on Theory and Applications of Satisfiability Testing*, pages 156–169, 2006.

41 A. Morgado, F. Heras, M. H. Liffiton, J. Planes, and J. Marques-Silva Iterative and core-guided MaxSAT solving: A survey and assessment. *Constraints*, 18(4):478–534, 2013.

42 D. Mitchell, B. Selman, and H. Leveque. A new method for solving hard satisfiability problems. In *Proceedings of the 10th National Conference on Artificial Intelligence (AAAI-92)*, pages 440–446, 1992.

43 B. Selman, H. A. Kautz, and B. Cohen. Noise strategies for improving local search. In B. Hayes-Roth and R. E. Korf, editors, *Proceedings of the 12th National Conference on Artificial Intelligence*, Seattle, WA, USA, July 31 - August 4, 1994, Volume 1, pages 337–343. AAAI Press / The MIT Press, 1994.

44 M. W. Moskewicz, C. F. Madigan, Y. Zhao, L. Zhang, and S. Malik. Chaff: Engineering an efficient SAT solver. In *Proceedings of the 38th Design Automation Conference, DAC 2001*, Las Vegas, NV, USA, June 18–22, 2001, pages 530–535. ACM, 2001.

45 J. P. Marques-Silva and K. A. Sakallah. GRASP: A search algorithm for propositional satisfiability. *IEEE Transactions on Computers*, 48(5):506–521, 1999.

46 S. Cai, K. Su, and A. Sattar. Local search with edge weighting and configuration checking heuristics for minimum vertex cover. *Artificial Intelligence*, 175(9-10):1672–1696, 2011.

47 L. Yang, X. Wang, H. Ding, Y. Yang, X. Zhao, and L. Pang. A survey of intelligent optimization algorithms for solving satisfiability problems. *Journal of Intelligent Fuzzy Systems*, 45(1):445–461, 2023.

48 M. Davis and H. Putnam. A computing procedure for quantification theory. *Journal of the ACM*, 7(3):201–215, 1960.

49 M. Davis, G. Logemann, and D. W. Loveland. A machine program for theorem-proving. *Communications of the ACM*, 5(7):394–397, 1962.

50 S. Buss and J. Nordström. Proof complexity and SAT solving. In A. Biere, M. Heule, H. van Maaren, and T. Walsh, editors, *Handbook of Satisfiability - Second Edition*, volume 336 of *Frontiers in Artificial Intelligence and Applications*, pages 233–350. IOS Press, 2021.

51 N. Eén and N. Sörensson. An extensible sat-solver. In *International Conference on Theory and Applications of Satisfiability Testing*, pages 502–518, Italy, May 2003. Springer.

52 G. Audemard, J.-M. Lagniez, and L. Simon. Improving glucose for incremental SAT solving with assumptions: Application to MUS extraction. In *Theory and Applications of Satisfiability Testing –SAT 2013*, pages 309–317, Berlin, Heidelberg, 2013. Springer-Verlag.

53 N. Manthey. The MergeSAT solver. In C.-M. Li and F. Manyà, editors, *Theory and Applications of Satisfiability Testing - SAT 2021 - 24th International Conference*, Barcelona, Spain, July 5–9, 2021, *Proceedings*, volume 12831 of *Lecture Notes in Computer Science*, pages 387–398. Springer, 2021.

54 M. Fleury and P. Lammich. A more pragmatic CDCL for IsaSAT and targetting LLVM (Short Paper). In B. Pientka and C. Tinelli, editors, *Automated Deduction - CADE 29 - 29th International Conference on Automated Deduction*, Rome, Italy, July 1–4, 2023, *Proceedings*, volume 14132 of *Lecture Notes in Computer Science*, pages 207–219. Springer, 2023.

55 S. Alouneh, S. Abed, M. H. Al Shayeji, and R. Mesleh. A comprehensive study and analysis on SAT-solvers: Advances, usages and achievements. *Artificial Intelligence Review*, 52(4):2575–2601, 2019.

56 C. M. Li and F. Manyà. MaxSAT, hard and soft constraints. In A. Biere, M. Heule, H. van Maaren, and T. Walsh, editors, *Handbook of Satisfiability - Second Edition*, volume 336 of *Frontiers in Artificial Intelligence and Applications*, pages 903–927. IOS Press, 2021.

57 D. S. Johnson. Approximation algorithms for combinatorial problems. *Journal of Computer and System Sciences*, 9(3):256–278, 1974.

58 M. X. Goemans and D. P. Williamson. Improved approximation algorithms for maximum cut and satisfiability problems using semidefinite programming. *Journal of the ACM*, 42(6):1115–1145, 1995.

59 J. Håstad. Some optimal inapproximability results. In *Proceedings of the 28th ACM Symposium on the Theory of Computing*, pages 1–10, 1997.

60 H. Karloff and U. Zwick. A 7/8-approximation algorithm for MAX 3SAT? In *Proceedings of the 38th Annual IEEE Symposium on Foundations of Computer Science*, pages 406–415, 1997.

61 M. F. Anjos. Semidefinite optimization approaches for satisfiability and maximum-satisfiability problems. *Journal on Satisfiability, Boolean Modeling and Computation*, 1(1):1–47, 2006.

62 F. Heras, J. Larrosa, and A. Oliveras. MiniMaxSat: An efficient weighted Max-SAT solver. *The Journal of Artificial Intelligence Research*, 31:1–32, 2008.

63 H. Lin, K. Su, and C. M. Li. Within-problem learning for efficient lower bound computation in Max-SAT solving. In *Proceedings of the 22th AAAI Conference on Artificial Intelligence*, pages 351–356, 2008.

64 K. Pipatsrisawat and A. Darwiche. Clone: Solving weighted Max-SAT in a reduced search space. In *Proceedings of 20th Australian Joint Conference on Artificial Intelligence*, pages 223–233, 2007.

65 T. Alsinet, F. Many, and J. Planes. An efficient solver for weighted Max-SAT. *Journal of Global Optimization*, 41(1):61–73, 2008.

66 J. Larrosa, F. Heras, and S. de Givry. A logical approach to efficient Max-SAT solving. *Artificial Intelligence*, 172(2-3):204–233, 2008.

67 J. Davies, J. Cho, and F. Bacchus. Using learnt clauses in MaxSAT. In *Proceedings of International Conference on Principles and Practice of Constraint Programming*, pages 176–190, 2010.

68 A. Kügel. Improved exact solver for the weighted MAX-SAT problem. In D. Le Berre, editor, *POS-10. Pragmatics of SAT*, Edinburgh, UK, July 10, 2010, volume 8 of *EPiC Series in Computing*, pages 15–27. EasyChair, 2010.

69 A. Abramé and D. Habet. Ahmaxsat: Description and evaluation of a branch and bound Max-SAT solver. *Journal on Satisfiability, Boolean Modeling and Computation*, 9(1):89–128, 2014.

70 F. Bacchus, M. Järvisalo, and R. Martins. Maximum satisfiabiliy. In A. Biere, M. Heule, H. van Maaren, and T. Walsh, editors, *Handbook of Satisfiability* - Second Edition, volume 336 of *Frontiers in Artificial Intelligence and Applications*, pages 929–991. IOS Press, 2021.

71 D. L. Berre and A. Parrain. The SAT4J library, release 2.2. *Journal on Satisfiability, Boolean Modeling and Computation*, 7:59–64, 2010.

72 M. Koshimura, T. Zhang, H. Fujita, and R. Hasegawa. QMaxSAT: A partial Max-SAT solver. *Journal on Satisfiability, Boolean Modeling and Computation*, 8:95–100, 2012.

73 T. Paxian, S. Reimer, and B. Becker. Pacose: An iterative SAT-based MaxSAT solver. *MaxSAT Evaluation 2018: Solver and Benchmark Descriptions*, B-2018-2 of Department of Computer Science Series of Publications B(1):20, 2018.

74 N. Manthey. Parallel by default - MergeSat and MergeSat-Pcasso. *MaxSAT Evaluation 2023: Solver and Benchmark Descriptions*, B-2023-1 of Department of Computer Science Series of Publications B(1):34–36, 2023.

75 Z. Fu and S. Malik. On solving the partial MAX-SAT problem. In A. Biere and C. P. Gomes, editors, *Theory and Applications of Satisfiability Testing - SAT 2006, 9th International Conference*, Seattle, WA, USA, August 12–15, 2006, *Proceedings*, volume 4121 of *Lecture Notes in Computer Science*, pages 252–265. Springer, 2006.

76 C. Ansótegui, M. L. Bonet, and J. Levy. Solving (weighted) partial MaxSAT through satisfiability testing. In *Proceedings of International conference on Theory and Applications of Satisfiability Testing*, pages 427–440, 2009.

77 V. M. Manquinho, J. Marques-Silva, and J. Planes. Algorithms for weighted Boolean optimization. In O. Kullmann, editor, *Theory and Applications of Satisfiability Testing - SAT 2009, 12th International Conference, SAT 2009*, Swansea, UK, June 30 - July 3, 2009. *Proceedings*, volume 5584 of *Lecture Notes in Computer Science*, pages 495–508. Springer, 2009.

78 A. Morgado, C. Dodaro, and J. Marques-Silva. Core-guided MaxSAT with soft cardinality constraints. In B. O'Sullivan, editor, *Principles and Practice of Constraint Programming - 20th International Conference, CP 2014*, Lyon, France, September 8–12, 2014. *Proceedings*, volume 8656 of *Lecture Notes in Computer Science*, pages 564–573. Springer, 2014.

79 M. Alviano. Maxino. *MaxSAT Evaluation 2017: Solver and Benchmark Descriptions*, B-2017-2 of Department of Computer Science Series of Publications B(1):10–11, 2017.

80 N. Narodytska and F. Bacchus. Maximum satisfiability using core-guided MaxSAT resolution. In C. E. Brodley and P. Stone, editors, *Proceedings of the 28th AAAI Conference on Artificial Intelligence*, July 27–31, 2014, Québec City, Québec, Canada, pages 2717–2723. AAAI Press, 2014.

81 J. Davies and F. Bacchus. Solving MAXSAT by solving a sequence of simpler SAT instances. In J. H.-M. Lee, editor, *Principles and Practice of Constraint Programming - CP 2011 - 17th International Conference, CP 2011*, Perugia, Italy, September 12–16, 2011. *Proceedings*, volume 6876 of *Lecture Notes in Computer Science*, pages 225–239. Springer, 2011.

82 J. Davies and F. Bacchus. Exploiting the power of MIP solvers in MaxSAT. In M. Järvisalo and A. Van Gelder, editors, *Theory and Applications of Satisfiability Testing - SAT 2013 - 16th International Conference*, Helsinki, Finland, July 8–12, 2013. *Proceedings*, volume 7962 of *Lecture Notes in Computer Science*, pages 166–181. Springer, 2013.

83 J. Davies and F. Bacchus. Postponing optimization to speed up MAXSAT solving. In C. Schulte, editor, *Principles and Practice of Constraint Programming - 19th International Conference, CP 2013*, Uppsala, Sweden, September 16–20, 2013. *Proceedings*, volume 8124 of *Lecture Notes in Computer Science*, pages 247–262. Springer, 2013.

84 P. Saikko, J. Berg, and M. Järvisalo. LMHS: A SAT-IP hybrid MaxSAT solver. In N. Creignou and D. Le Berre, editors, *Theory and Applications of Satisfiability Testing - SAT 2016 - 19th International Conference*, Bordeaux, France, July 5–8, 2016, *Proceedings*, volume 9710 of *Lecture Notes in Computer Science*, pages 539–546. Springer, 2016.

85 R. Martins, V. M. Manquinho, and I. Lynce. Exploiting cardinality encodings in parallel maximum satisfiability. In *IEEE 23rd International Conference on Tools with Artificial Intelligence, ICTAI 2011*, Boca Raton, FL, USA, November 7–9, 2011, pages 313–320. IEEE Computer Society, 2011.

86 M. Terra-Neves, I. Lynce, and V. M. Manquinho. Non-portfolio approaches for distributed maximum satisfiability. In *28th IEEE International Conference on Tools with Artificial Intelligence, ICTAI 2016*, San Jose, CA, USA, November 6–8, 2016, pages 436–443. IEEE Computer Society, 2016.

87 A. Biere. Lingeling essentials, A tutorial on design and implementation aspects of the SAT solver lingering. In D. Le Berre, editor, *POS-14. 5th Pragmatics of SAT Workshop, A Workshop of the SAT 2014 Conference, Part of FLoC 2014 During the Vienna Summer of Logic*, July 13, 2014, Vienna, Austria, volume 27 of *EPiC Series in Computing*, page 88. EasyChair, 2014.

88 H. Nabeshima and K. Inoue. Reproducible efficient parallel SAT solving. In L. Pulina and M. Seidl, editors, *Theory and Applications of Satisfiability Testing - SAT 2020 - 23rd International Conference*, Alghero, Italy, July 3–10, 2020, *Proceedings*, volume 12178 of *Lecture Notes in Computer Science*, pages 123–138. Springer, 2020.

89 J. Zheng, K. He, J. Zhou, Y. Jin, C.-M. Li, and F. Manyà. BandMaxSAT: A local search MaxSAT solver with multi-armed bandit. *arXiv preprint arXiv:2201.05544*, 2022.

90 H. Wu. Improving Sat-solving with machine learning. In M. E. Caspersen, S. H. Edwards, T. Barnes, and D. D. Garcia, editors, *Proceedings of the 2017 ACM SIGCSE Technical Symposium on Computer Science Education, SIGCSE 2017*, Seattle, WA, USA, March 8–11, 2017, pages 787–788. ACM, 2017.

91 W. Guo, H.-L. Zhen, X. Li, W. Luo, M. Yuan, Y. Jin, and J. Yan. Machine learning methods in solving the Boolean satisfiability problem. *Machine Intelligence Research*, 20(5):640–655, 2023.

92 M. Liu, P. Huang, F. Jia, F. Zhang, Y. Sun, S. Cai, F. Ma, and J. Zhang. Can graph neural networks learn to solve the MaxSAT problem? (Student

Abstract). *Proceedings of the AAAI Conference on Artificial Intelligence*, 37(13):16264–16265, Sept. 2023.

93 C. M. Varmantchaonala, J. L. E. K. Fendji, J. P. T. Njafa, and M. Atemkeng. Quantum hybrid algorithm for solving SAT problem. *Engineering Applications of Artificial Intelligence*, 121:106058, 2023.

94 S.-W. Lin, T.-F. Wang, Y.-R. Chen, Z. Hou, D. Sanán, and Y. S. Teo. A parallel and distributed quantum SAT solver based on entanglement and quantum teleportation. *CoRR*, abs/2308.03344, 2023.

95 A. Alasow and M. Perkowski. Quantum algorithm for maximum satisfiability. In *2022 IEEE 52nd International Symposium on Multiple-Valued Logic (ISMVL)*, pages 27–34, 2022.

2
Encoding in General

After reading this chapter, you should be able to:

- Explore the process of transforming a propositional formula to a logically equivalent formula in conjunctive normal form (CNF) using Boolean algebra and Tseitin encoding.
- Gain knowledge about commonly used satisfiability problem-solving environments, including the DIMACS format and the PySAT Python toolkit.
- Learn the general problem encoding and solving process through a concrete example.

2.1 CNF Encodings

Before a combinational optimization problem can be solved by satisfiability (SAT) or maximum satisfiability (MaxSAT) solvers, it must usually be transformed to conjunctive normal form (CNF), which is a very simple form, leading to easy algorithm implementation and a common file format. However, after being transformed to CNF, the formula may lose a great deal of structural information. To conquer this drawback, solvers for non-CNF problems have been devised [1–4]. These techniques can yield great improvements on certain structured problems but CNF currently remains the norm [5]. This section describes techniques for transforming from a propositional formula to CNF, including transformation by Boolean algebra and Tseitin encoding. A propositional formula is a formal expression that denotes a proposition, which is a statement telling either true or false.

Applied Satisfiability: Cryptography, Scheduling, and Coalitional Games,
First Edition. Xiaojuan Liao and Miyuki Koshimura.

2.1.1 Transformation by Boolean Algebra

Boolean algebra, introduced by George Boole, is a subarea of algebra in which the values of the variables are the truth values *true* (written as 1) and *false* (written as 0). The basic operations in Boolean algebra are conjunction (*and*), disjunction (*or*), and negation (*not*), denoted as $\wedge$, $\vee$, $\neg$, respectively.

A propositional formula can be transformed to a logically equivalent CNF formula by using laws of Boolean algebra. A law of Boolean algebra is an identity between two Boolean terms, where a Boolean term is defined as an expression built up from variables and the constants 0 and 1 using the basic operations $\wedge$, $\vee$, $\neg$. Important properties of Boolean algebra are monotone and nonmonotone laws. Both are typically helpful in simplifying and converting Boolean formulas to CNF.

Monotone laws are associated with conjunction and disjunction. These operations have the property that changing either argument either keeps the output unchanged or causes the output to change in the same way as the input. In other words, changing any variable from 0 to 1 never results in the output changing from 1 to 0. Operations with this property are referred to as monotone. Table 2.1 presents monotone laws that hold in Boolean algebra, with x, y, and z representing Boolean variables. Nonmonotonicity, on the other hand, is introduced via complement $\neg$ [6]. Nonmonotone laws are listed in Table 2.2.

Table 2.1 Monotone laws of Boolean algebra.

Axiom	Description
Associativity of $\vee$	$x \vee (y \vee z) = (x \vee y) \vee z$
Associativity of $\wedge$	$x \wedge (y \wedge z) = (x \wedge y) \wedge z$
Commutativity of $\vee$	$x \vee y = y \vee x$
Commutativity of $\wedge$	$x \wedge y = y \wedge x$
Distributivity of $\wedge$ over $\vee$	$x \wedge (y \vee z) = (x \wedge y) \vee (x \wedge z)$
Identity for $\vee$	$x \vee 0 = x$
Identity for $\wedge$	$x \wedge 1 = x$
Annihilator for $\wedge$	$x \wedge 0 = 0$
Annihilator for $\vee$	$x \vee 1 = 1$
Idempotence of $\vee$	$x \vee x = x$
Idempotence of $\wedge$	$x \wedge x = x$
Absorption 1	$x \wedge (x \vee y) = x$
Absorption 2	$x \vee (x \wedge y) = x$
Distributivity of $\vee$ over $\wedge$	$x \vee (y \wedge z) = (x \vee y) \wedge (x \vee z)$

Table 2.2 Nonmonotone laws of Boolean algebra.

Axiom	**Description**
Complementation law 1	$x \wedge \neg x = 0$
Complementation law 2	$x \vee \neg x = 1$
Double negation	$\neg(\neg x) = x$
De Morgan's law 1	$\neg x \wedge \neg y = \neg(x \vee y)$
De Morgan's law 2	$\neg x \vee \neg y = \neg(x \wedge y)$

Basic operations, i.e., $\wedge$, $\vee$, $\neg$, can be taken as a basis for other derived Boolean operations that can be built up from them by *composition*, the manner in which operations are combined or compounded. Some examples of derived operations composed from the basic operations are shown as follows.

$$x \to y = \neg x \vee y$$

$$x \oplus y = (x \vee y) \wedge \neg (x \wedge y)$$

$$x \equiv y = \neg (x \oplus y)$$

The operation $x \to y$ is called *material implication*. If x is true, then the value of $x \to y$ is taken to be that of y. However, if x is false, the value of $x \to y$ is constantly true regardless of that of y. $x \oplus y$ is called *exclusive or*. It excludes the possibility of x and y taking the same truth value. In other words, the value of $x \oplus y$ is true just when x and y have different truth values. $x \equiv y$, the complement of exclusive or, is *equivalence* or *Boolean equality*. The value of $x \equiv y$ is true just when x and y have the same value.

Finally, we take an example [2] of a propositional formula to show how it is transformed to a CNF formula by Boolean algebra.

Example 2.1 Let us consider the following propositional formula, in which all letters represent Boolean variables:

$$(a \to (c \wedge d)) \vee (b \to (c \wedge e))$$

The implications can be decomposed:

$$((a \to c) \wedge (a \to d)) \vee ((b \to c) \wedge (b \to e))$$

The conjunctions and disjunctions can be rearranged:

$$((a \to c) \vee (b \to c)) \wedge ((a \to c) \vee (b \to e)) \wedge ((a \to d) \vee (b \to c)) \wedge ((a \to d) \vee (b \to e))$$

The implications can be rewritten as disjunctions, and duplicated literals are removed:

$$(\neg a \vee \neg b \vee c) \wedge (\neg a \vee \neg b \vee c \vee e) \wedge (\neg a \vee \neg b \vee c \vee d) \wedge (\neg a \vee \neg b \vee d \vee e)$$

Here, we notice that the clause $(\neg a \vee \neg b \vee c)$ *subsumes* the clause $(\neg a \vee \neg b \vee c \vee e)$. In other words, if an assignment satisfies the former clause, it will surely satisfy the latter one. In this case, the subsumed clause $(\neg a \vee \neg b \vee c \vee e)$ can be removed. Likewise, the clause $(\neg a \vee \neg b \vee c)$ subsumes the clause $(\neg a \vee \neg b \vee c \vee d)$, allowing for the removal of $(\neg a \vee \neg b \vee c \vee d)$. After removing the subsumed clauses, the conjunction is left as follow:

$$(\neg a \vee \neg b \vee c) \wedge (\neg a \vee \neg b \vee d \vee e)$$

This propositional formula finally reduces to a compact CNF formula, but in general, this method generates exponentially large formulas [5].

2.1.2 Transformation by Tseitin Encoding

The naive approach to use rules of Boolean algebra results in an exponential increase in equation size. To avoid the exponentially large size, conversion to CNF is usually achieved by using the well-known Tseitin encoding [7], which outputs an equation whose size grows linearly relative to the input.

Tseitin encoding generates a new variable (or say, auxiliary variable), for each subformula of the original formula, as well as a set of clauses that capture the equivalent relationship[1] between these new variables and the subformulas. The procedure of Tseitin encodings is exhibited in the following example [5].

Example 2.2 Consider the propositional formula in Example 2.1. A new Boolean variable f_1 is introduced to replace the subformula $(c \wedge d)$, with the definition:

$$f_1 \leftrightarrow (c \wedge d)$$

This definition is reduced to clauses:

$$(\neg f_1 \vee c) \wedge (\neg f_1 \vee d) \wedge (\neg c \vee \neg d \vee f_1)$$

Similarly, another new Boolean variable f_2 is introduced to replace the subformula $(c \wedge e)$, with the definition:

$$f_2 \leftrightarrow (c \wedge e)$$

1 Formulas ϕ and ϕ' are equivalent means ϕ is satisfiable if and only if ϕ' is satisfiable.

which is reduced to clauses

$$(\neg f_2 \vee c) \wedge (\neg f_2 \vee e) \wedge (\neg c \vee \neg e \vee f_2)$$

Then the original formula is reduced to:

$$(\neg a \vee f_1) \wedge (\neg b \vee f_2)$$

Next, two more variables are introduced with definitions:

$$f_3 \leftrightarrow (\neg a \vee f_1)$$

and

$$f_4 \leftrightarrow (\neg b \vee f_2)$$

and clauses:

$$(\neg f\ (\neg f_3 \vee \neg a \vee f_1) \wedge (a \vee f_3) \wedge (\neg f_1 \vee f_3) \wedge (\neg f_4 \vee \neg b \vee f_2) \wedge (b \vee f_4) \wedge (\neg f_2 \vee f_4))$$

The formula is now reduced to:

$$(f_3 \vee f_4)$$

This formula is already in clausal form as we desired.

Tseitin encoding is linear in the size of the original formula as long as the Boolean operators that appear in the formula have linear clausal encodings. The operators *and*, *or*, *not*, and *implies* all have linear clausal encodings, while encoding *exclusive or* requires exponential number of clauses to that of variables. We will introduce the CNF encoding of *exclusive or* when necessary in Chapter 4.

2.2 Satisfiability Problem-Solving Environments

SAT and MaxSAT solving techniques keep on making progress. To keep up the driving force in enhancing SAT and MaxSAT solvers, SAT competitions (http://satcompetition.org, starting from 2002) and MaxSAT evaluations (https://maxsat-evaluations.github.io, starting from 2006) provide opportunities for solver developers to present their works to objectively compare the performance of their solvers with that of other state-of-the-art solvers. All benchmark instances used in evaluations follow the same input format, which will be introduced in Section 2.2.1. Additionally, the power of SAT technology has been integrated into a Python toolkit called PySAT, which includes a variety of widely used solvers, cardinality encodings, and pseudo-Boolean encodings. The usage of PySAT will be introduced in Section 2.2.2.

2.2.1 DIMACS Format

SAT competitions and MaxSAT evaluations provide a collection of problem instances for solvers to tackle. This subsection describes the file format for submitting instances for evaluation.

The benchmark file is a simplified version of the DIMACS format, typically with extensions *.cnf* for SAT and MaxSAT problems and *.wcnf* for extended MaxSAT problems, i.e., weighted MaxSAT, partial MaxSAT, and weighted partial MaxSAT.

A .cnf file usually starts with comment lines, beginning with the character "c." Right after the comments, there is a line summarizing the parameter information of the instance. The standard form is "p cnf nbvar nbclauses," indicating that the instance is in CNF format, and the exact numbers of variables and clauses in the file are "nbvar" and "nbclauses," respectively. Then the clauses follow. Each clause is a sequence of distinct non-null numbers between "-nbvar" and "nbvar" ending with 0 in each line. Positive numbers denote the corresponding variables, and negative numbers denote the negations of the corresponding variables. An exemplified .cnf file is shown in Example 2.3.

Example 2.3 A CNF formula $\phi = \{x_1 \vee \neg x_5 \vee x_4, \neg x_2 \vee x_5 \vee x_3 \vee x_4, \neg x_3 \vee \neg x_4\}$ is written in the following .cnf file in DIMACS format.

```
c
c start with comments
c
p cnf 5 3
1 −5 4 0
−2 5 3 4 0
− 3 −4 0
```

In this example, "p cnf 5 3" is the parameters line, indicating that the CNF instance contains 5 variables and 3 clauses. Following the parameter line, there are three clauses listed line by line. Taking the first clause line as an example, "1 −5 4 0" represents the clause $x_1 \vee \neg x_5 \vee x_4$, where x_1, x_5, and x_4 are Boolean variables denoted by integers 1, 5, and 4 in the file, respectively.

The input file can be read by the command *./solvername ⟨input_file_name⟩*. A SAT solver tells whether the input instance is satisfiable, and a MaxSAT solver returns the maximum number of satisfied clauses in the instance. For solving extensions of MaxSAT, minor changes should be made on the DIMACS format.

In weighted MaxSAT, the parameters line is "p wcnf nbvar nbclauses," indicating that the instance is in wcnf format. The weight of each clause, indicated by the first integer in each clause line, is an integer greater than or equal to 1, and the sum of all clauses must be smaller than 2^{63}. A .wcnf file representing a weighted MaxSAT problem is shown as Example 2.4.

Example 2.4 A weighted MaxSAT formula $\phi = \{(x_1 \vee \neg x_2, 10), (\neg x_1 \vee x_2 \vee \neg x_3, 3), (\neg x_3 \vee x_2, 8), (x_1 \vee x_3, 5)\}$ is written in the following .wcnf file in DIMACS format.

```
c
c weighted MaxSAT comments
c
p wcnf 3 4
10 1 -2 0
3 -1 2 -3 0
8 -3 2 0
5 1 3 0
```

In partial MaxSAT, the parameters line is "p wcnf nbvar nbclauses top," followed by clauses listed line by line. Each clause is associated with a weight, which is the first integer in the clause. Hard clauses have weights equal or greater than "top" that is declared in the parameters line. Soft clauses have weight 1 and the sum of weights of all soft clauses must be smaller than 2^{63}. It must be assured that "top" is a weight always greater than the sum of the weights of violated soft clauses in the solution. A .wcnf file representing a partial MaxSAT problem is shown as Example 2.5.

Example 2.5 A partial MaxSAT formula $\phi = \{[x_1 \vee \neg x_2 \vee x_4], [\neg x_1 \vee \neg x_2 \vee x_3], (\neg x_2 \vee \neg x_4), (\neg x_3 \vee x_2), (x_1 \vee x_3)\}$ is written in the following .wcnf file in DIMACS format.

```
c
c partial MaxSAT comments
c
p wcnf 4 5 15
15 1 -2 4 0
1 -1 -2 3 0
1 -2 -4 0
1 -3 2 0
1 1 3 0
```

Weighted partial MaxSAT is the combination of weighted MaxSAT and partial MaxSAT. Correspondingly, the .wcnf file is the mixed form of the two. The parameters line is "p wcnf nbvar nbclauses top," followed by clause lines. Each clause is associated with a weight, which is the first integer in the clause. Hard clauses have weight equal or greater than "top" and soft clauses have a weight smaller than "top." The sum of weights of all soft clauses is smaller than 2^{63}. It must be assured that "top" is a weight always greater than the sum of the weights of violated soft clauses in the solution. In practice, the value of "top" is set exactly one larger than the sum of weights of all soft clauses in the problem. A .wcnf file representing a weighted partial MaxSAT problem is shown as Example 2.6.

Example 2.6 A weighted partial MaxSAT formula $\phi = \{[x_1 \vee \neg x_2 \vee x_4], [\neg x_1 \vee \neg x_2 \vee x_3], (\neg x_2 \vee \neg x_4, 8), (\neg x_3 \vee x_2, 4), (x_1 \vee x_3, 3)\}$ is written in the following.wcnf file in DIMACS format.

```
c
c weighted partial MaxSAT comments
c
p wcnf 4 5 16
16 1 −2 4 0
16 −1 −2 3 0
8 −2 −4 0
4 −3 2 0
3 1 3 0
```

2.2.2 PySAT: Python Toolkit

Recently, a simple and unified incremental interface to a number of SAT and MaxSAT solvers is integrated in a Python toolkit called PySAT [8], which further facilitates the utilization of solvers. By exploiting PySAT, one can easily realize automatic conversion from regular arithmetic expression to CNF formulas and call the solver that is embedded in the toolkit.

The PySAT toolkit has five core modules: card, formula, process, pb, and solvers, providing access to cardinality encodings, Boolean formula manipulation, preprocessor functionality of CaDiCaL 1.5.3, Pseudo-Boolean encodings, and SAT solvers' application programming interface (API), respectively. This subsection introduces core class methods in these modules that will be used in the rest of this book. Readers may refer to PySAT documentation[2] for complete information.

Class for manipulating CNF formulas is "pysat.formula.CNF," which provides a sort of methods of creating, reading, and writing formulas. Example 2.7 describes how to create a CNF formula with PySAT.

---

2 https://pysathq.github.io/docs/html/index.html\LY1\textbackslash#api-documentation.

Example 2.7 A CNF formula $\phi = \{x_1 \vee \neg x_5 \vee x_4, \neg x_2 \vee x_5 \vee x_3 \vee x_4, \neg x_3 \vee \neg x_4\}$ (the same as that in Example 2.3) can be created by the following piece of code:

```
from pysat.formula import CNF
cnf = CNF() # create an object of class CNF
cnf.append([1,-5,4])  # add a clause to cnf
cnf.append([-2,5,3,4]) # add a clause to cnf
cnf.append([-3,-4]) # add a clause to cnf
```

Alternatively, the following piece of code fulfills the same functionality:

```
from pysat.formula import CNF
cnf = CNF() # create an object of class CNF
# add several clauses to a CNF formula
cnf.extend([[1,-5,4], [-2,5,3,4], [-3,-4]])
```

Class for manipulating (weighted) partial CNF formulas is "pysat.formula. WCNF." Example 2.8 shows how to create a weighted partial CNF formula with PySAT.

Example 2.8 A Weighted Conjunctive Normal Form (WCNF) formula $\phi = \{[x_1 \vee \neg x_2 \vee x_4], [\neg x_1 \vee \neg x_2 \vee x_3], (\neg x_2 \vee \neg x_4, 8), (\neg x_3 \vee x_2, 4), (x_1 \vee x_3, 3)\}$ (the same as that in Example 2.6) can be created and read by the following piece of code:

```
from pysat.formula import WCNF
wcnf = WCNF() # create an object of class WCNF
wcnf.append([1,-2,4]) # add a hard clause
wcnf.append([-1,-2,3]) # add a hard clause
wcnf.append([-2,-4], weight=8) # add a weighted soft clause
wcnf.append([-3,2], weight=4) # add a weighted soft clause
wcnf.append([1,3], weight=3) # add a weighted soft clause
```

Hard clauses, soft clauses, and weights can be read separately:

```
print(wcnf.hard) # [[1,-2,4],[-1,-2,3]]
print(wcnf.soft) # [[-2,-4],[-3,2],[1,3]]
print(wcnf.wght) # [8,4,3]
```

The abstract class "pysat.card.CardEnc" is responsible for encoding cardinality constraints to a CNF formula. Specifically, it has three class methods

"CardEnc.atleast," "CardEnc.atmost," and "CardEnc.equals" to encode AtLeastK constraint (i.e., $\sum_{i=1}^{n} x_i \geq k$), AtMostK constraint (i.e., $\sum_{i=1}^{n} x_i \leq k$), and EqualsK constraint (i.e., $\sum_{i=1}^{n} x_i = k$) into CNF formulas. Each of these methods returns an object of class "pysat.formula.CNFPlus" representing the resulting CNF formula. Class CNFPlus inherits most of the functionality of the CNF class. For simplicity, we do not distinguish them here. Example 2.9 shows the way of encoding an AtMostK constraint with PySAT. AtLeastK and EqualsK constraints can be dealt with in a similar way. They all share the arguments and the return type.

Example 2.9 A cardinality constraint $x_1 + x_2 + x_3 \leq 1$ can be encoded to a CNF formula by using the class method "CardEnc.atmost," shown in the following piece of code:

```
from pysat.card import *
cnf=CardEnc.atmost(lits=[1,2,3],bound=1,encoding=0)
print(cnf.clauses) # [[-1,-2], [-1,-3], [-2,-3]]
```

Key arguments in CardEnc.atmost are described as follows: "lits" gives a list of literals in the sum and "bound" gives the value of bound k. The argument "encoding" indicates the identifier of the used encoding. Specifically, "encoding = 0" signifies that the desired encoding is pairwise. For complete information on values denoting the encodings, readers may refer to the documentation of pysat.card.[3] Eventually, the cardinality constraint is converted to a CNF formula $\{\neg x_1 \vee \neg x_2, \neg x_1 \vee \neg x_3, \neg x_2 \vee \neg x_3\}$.

The "pysat.pb" module provides access to pseudo-Boolean encodings. A pseudo-Boolean constraint is a constraint of the form: $\sum_{i=1}^{n} a_i x_i \{\geq, \leq, =\} k$, where a_i and k are integers and x_i are literals. Functionality of this module is available only if the PyPBLib package is installed. The following piece of code installs PyPBLib from PyPI:

```
$ pip install pypblib
```

The abstract class "pysat.pb.PBEnc" is responsible for encoding pseudo-Boolean constraints to a CNF formula. Specifically, it has three class methods "PBEnc.atleast," "PBEnc.atmost," and "PBEnc.equals" to encode Greater than or Equal (GEQ) constraint (i.e., $\sum_{i=1}^{n} a_i x_i \geq k$), Less than or Equal (LEQ) constraint (i.e., $\sum_{i=1}^{n} a_i x_i \leq k$), and Equals constraint (i.e., $\sum_{i=1}^{n} a_i x_i = k$) into CNF formulas. Each of these methods returns an object of class "pysat.formula.CNF"

3 https://pysathq.github.io/docs/html/api/card.html.

representing the resulting CNF formula. Example 2.10 shows the way of encoding a LEQ constraint with PySAT. GEQ and Equals constraints can be dealt with in a similar manner. They all share the arguments and the return type.

Example 2.10 A pseudo-Boolean constraint $x_1 + 2 \cdot x_2 + 3 \cdot x_3 \leq 3$ can be encoded to a CNF formula by using the class method "PBEnc.atmost," shown in the following piece of code:

```
from pysat.pb import *
cnf=PBEnc.atmost(lits=[1,2,3],weights=[1,2,3],bound=3)
print(cnf.clauses) # [[4],[-1,-5],[-2,-5],[5,-3,-6],[6]]
```

Key arguments in PBEnc.atmost are described as follows: "lits" provides a list of literals in the sum; "weights" provides a list of weights; "bound" specifies the value of bound k. The argument "encoding" indicates the identifier of the used encoding. In this example, "encoding" takes the default value, i.e., 0, and thus is omitted in the declaration. Values denoting the encodings are exhibited in the documentation of pysat.pb.[4] Eventually, the cardinality constraint is converted to a CNF formula $\{x_4, \neg x_1 \vee \neg x_5, \neg x_2 \vee \neg x_5, x_5 \vee \neg x_3 \vee \neg x_6, x_6\}$.

The "pysat.solvers" module provides incremental access to modern SAT solvers. Currently, solvers supported by PySAT include CaDiCaL, Glucose (versions 3.0 and 4.1), Lingeling, MapleLCMDistChronoBT, MapleCM, Maplesat, Mergesat (version 3.0), Minicard (version 1.2), Minisat (version 2.2 and GitHub version), etc. Example 2.11 shows how to solve a CNF formula (the same as that in Example 2.3) with an off-the-shelf solver. Readers may refer to SAT solver' API[5] to access all the PySAT supported solvers.

Example 2.11 The following piece of code generates a SAT problem instance and solve it with Glucose 3.0.

```
from pysat.solvers import Glucose3
s = Glucose3() # create an object of Glucose3
s.add_clause([1,-5,4]) # add a clause to s
s.add_clause([-2,5,3,4]) # add a clause to s
s.add_clause([-3,-4]) # add a clause to s
print(s.solve()) # True
```

4 https://pysathq.github.io/docs/html/api/pb.html.
5 https://pysathq.github.io/docs/html/api/solvers.html.

Alternatively, the following piece of code realizes the same functionality:

```
from pysat.formula import CNF
from pysat.solvers import Glucose3
cnf = CNF() # create an object of class CNF
# add several clauses to a CNF formula
cnf.extend([[1,-5,4], [-2,5,3,4], [-3,-4]])
with Glucose3(bootstrap_with=cnf.clauses) as s:
    print(s.solve()) # True
```

The argument "bootstrap_with" accepts a list of clauses for solver initialization. It is particularly suitable for cases where batches of clauses are generated in advance.

In this example, the method "s.solve()" check satisfiability and returns True, indicating that the input clauses are all satisfied. Since the input formula is satisfiable, the feasible assignments of variables can be obtained using the following code:

```
print(s.get_model()) # [-1, -2, -3, -4, -5]
```

which indicates that the returned feasible assignments are $x_1 = 0, x_2 = 0, x_3 = 0, x_4 = 0, x_5 = 0$.

In addition to its core modules, PySAT provides supplementary module examples packages to demonstrate practical problem-solving applications of the toolkit. MaxSAT solvers are included in the packages. As an illustration, we will use the LSU algorithm for MaxSAT (pysat.examples.lsu) to showcase the use of the toolkit in solving MaxSAT problems.

The module implements a prototype of the known LSU/LSUS, e.g., linear (search) satisfiable (SAT)-unsatisfiable (UNSAT), algorithm for MaxSAT [9]. The algorithm can be accessed and invoked through the standard import interface of Python. Example 2.12 shows the partial MaxSAT problem generation and solving process.

Example 2.12 The following piece of code generates a partial MaxSAT problem instance:

```
from pysat.formula import WCNF
from pysat.examples.lsu import LSU
wcnf = WCNF() # create an object of class WCNF
wcnf.append([-1,-2]) # add a hard clause
wcnf.append([-1,-3]) # add a hard clause
```

```
wcnf.append([-2,-3]) # add a hard clause
wcnf.append([1], weight=1) # add a soft clause
wcnf.append([2], weight=1) # add a soft clause
wcnf.append([3], weight=1) # add a soft clause
```

Up to this point, a partial MaxSAT problem instance has been stored in the object wcnf: $\phi = \{[\neg x_1 \vee \neg x_2], [\neg x_1 \vee \neg x_3], [\neg x_2 \vee \neg x_3], (x_1), (x_2), (x_3)\}$. The following piece of code solves the instance and returns the solving result. Specifically, "LSU(wcnf)" creates an object of the LSU class, taking wcnf as the input. The method "s.solve()" computes a solution to the problem, returning True if all the hard clauses are satisfied and False otherwise. The attributes "s.cost" and "s.model" provide the number of the falsified clauses and the feasible assignment leading to the cost, respectively.

```
s = LSU(wcnf) # input wcnf to an object of LSU
print(s.solve()) # True
print(s.cost) # 2
print(list(s.model)) # [-1, -2, 3]
```

2.3 Case Study

This section takes the well-known N-queens problem as an example to illustrate the process of encoding and problem solving. Consider a square chessboard, which is divided into $n \times n$ smaller squares (cells). The problem is to place n queens (chess pieces) on the $n \times n$ chessboard so that no two queens can attack each other, that is, no two queens are placed on the same row, column or diagonal. Figure 2.1 shows two feasible solutions for 4-queens.

There are a large number of ways of modeling the N-queens problem as SAT. Typically, Nadel [10] presents no fewer than nine constraint models. We exhibit the most obvious one to show how a SAT solver is utilized to get the solution.

For each position on the chessboard, define a Boolean variable: P_{ij} $(1 \leq i \leq n, 1 \leq j \leq n)$, which is true if a cell on the ith row and jth column is placed a queen. Then, add constraints to prevent attacks:

(S1) $\forall 1 \leq i \leq n$, there is only one queen on the ith row:

$$\sum_{j=1}^{n} P_{ij} = 1 \tag{2.1}$$

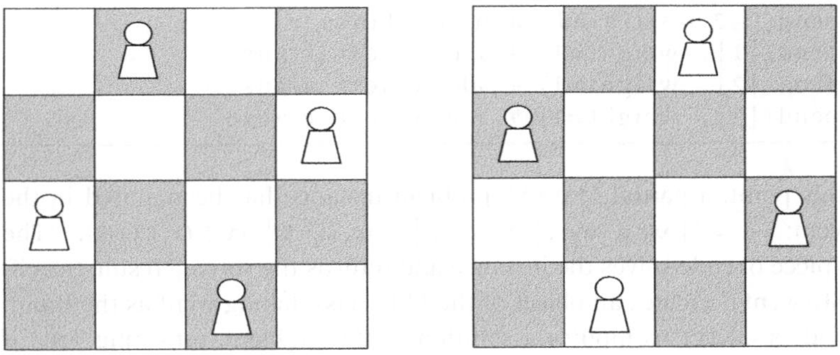

(a) (b)

Figure 2.1 Solutions of the 4-queens problem. (a) Solution 1 and (b) Solution 2 (Source: Xiaojuan Liao and Miyuki Koshimura/original).

(S2) $\forall 1 \leq j \leq n$, there is only one queen on the jth column:

$$\sum_{i=1}^{n} P_{ij} = 1 \tag{2.2}$$

(S3) No two queens are on the same diagonal (from top left to bottom right):

$$\sum_{\substack{i-j=d\\1\leq i\leq n\\1\leq j\leq n}} P_{i,j} \leq 1 \quad (-n+2 \leq d \leq n-2) \tag{2.3}$$

(S4) No two queens are on the same antidiagonal (from bottom left to top right):

$$\sum_{\substack{i+j=d\\1\leq i\leq n\\1\leq j\leq n}} P_{i,j} \leq 1 \quad (3 \leq d \leq 2n-1) \tag{2.4}$$

Until now, the N-queens problem has been transformed into Boolean formulas. Through manual encoding or utilizing the PySAT toolkit, these formulas can be translated into a set of clauses, and an off-the-shelf SAT solver can determine whether the formula is satisfiable. If the formula is satisfiable, a feasible solution to the problem can be obtained from the corresponding variable assignment. If the formula is unsatisfiable, it indicates that no solution exists for the problem. For example, the 4-queens problem is satisfiable, whereas the 2-queens and 3-queens problems are unsatisfiable.

Example 2.13 Consider the 4-queens problem. We define Boolean variables P_{ij} for $1 \leq i \leq 4$ and $1 \leq j \leq 4$ to indicate whether each cell is placed a queen.

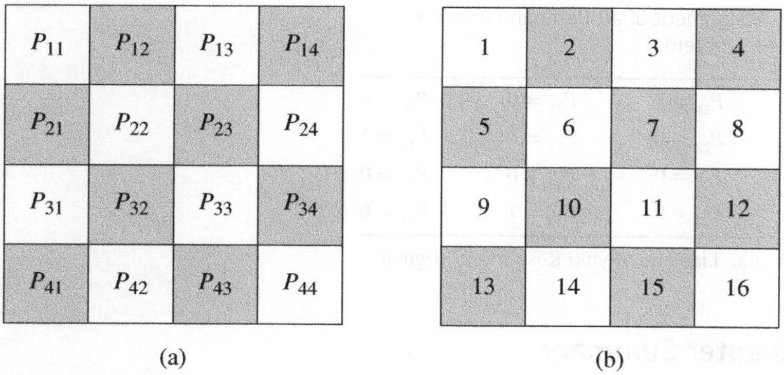

Figure 2.2 Variable representation of the 4-queens problem. (a) Variable notation and (b) variable numbering (Source: Xiaojuan Liao and Miyuki Koshimura/original).

Table 2.3 Constraints of the 4-queens problem.

(S1)	$P_{11}+P_{12}+P_{13}+P_{14}=1$	(S2)	$P_{11}+P_{21}+P_{31}+P_{41}=1$
	$P_{21}+P_{22}+P_{23}+P_{24}=1$		$P_{12}+P_{22}+P_{32}+P_{42}=1$
	$P_{31}+P_{32}+P_{33}+P_{34}=1$		$P_{13}+P_{23}+P_{33}+P_{43}=1$
	$P_{41}+P_{42}+P_{43}+P_{44}=1$		$P_{14}+P_{24}+P_{34}+P_{44}=1$
(S3)	$P_{13}+P_{24}\leq 1$	(S4)	$P_{12}+P_{21}\leq 1$
	$P_{12}+P_{23}+P_{34}\leq 1$		$P_{13}+P_{22}+P_{31}\leq 1$
	$P_{11}+P_{22}+P_{33}+P_{44}\leq 1$		$P_{14}+P_{23}+P_{32}+P_{41}\leq 1$
	$P_{21}+P_{32}+P_{43}\leq 1$		$P_{24}+P_{33}+P_{42}\leq 1$
	$P_{31}+P_{42}\leq 1$		$P_{34}+P_{43}\leq 1$

Source: Xiaojuan Liao et al./Original.

The variable notation is exhibited in Figure 2.2, and the constraints are shown Table 2.3.

These Boolean formulas can be translated into a standard SAT problem through manual encoding or by using the PySAT toolkit. The complete file representing the SAT problem in DIMACS CNF input format is provided in Appendix A.1. For clarity, Figure 2.2b displays the variable numbering as it appears in the DIMACS-formatted file.

Taking the DIMACS CNF file as input, any SAT solver can be exploited to find a feasible assignment if the formula is satisfiable. Table 2.4 lists the assigned truth values of all the Boolean variables in the 4-queens problem, which correspond to the solution 2 as illustrated in Figure 2.1.

Table 2.4 Assignment of all Boolean variables in the 4-queens problem.

$P_{11} = 0$	$P_{21} = 1$	$P_{31} = 0$	$P_{41} = 0$
$P_{12} = 0$	$P_{22} = 0$	$P_{32} = 0$	$P_{42} = 1$
$P_{13} = 1$	$P_{23} = 0$	$P_{33} = 0$	$P_{43} = 0$
$P_{14} = 0$	$P_{24} = 0$	$P_{34} = 1$	$P_{44} = 0$

Source: Xiaojuan Liao and Miyuki Koshimura/original.

2.4 Chapter Summary

This chapter reviewed both theoretical and empirical works on encoding. Currently, CNF is widely accepted as the standard format for inputting into SAT or MaxSAT solvers. This chapter first covered techniques of transforming a propositional Boolean formula into CNF using Boolean algebra and Tseitin encodings. Subsequently, environments for solving satisfiability problems were introduced, including the DIMACS file format for CNF and the PySAT Python toolkit. Finally, a concrete example involving the N-queens problem was presented, demonstrating how to encode the problem in CNF and utilize off-the-shelf solvers to obtain solutions.

References

1 Z. Stachniak. Going non-clausal. In *Proceedings of the 5th International Symposium on Theory and Applications of Satisfiability Testing*, pages 316–322, 2002.

2 C. Thiffault, F. Bacchus, and T. Walsh. Solving non-clausal formulas with DPLL search. In *Proceedings of the 7th International Conference on Theory and Applications of Satisfiability Testing*, pages 147–156, 2004.

3 R. Muhammad and P. J. Stuckey. A stochastic non-CNF SAT solver. In *Proceedings of Trends in Artificial Intelligence, 9th Pacific Rim International Conference on Artificial Intelligence*, pages 120–129, 2006.

4 D. N. Pham, J. R. Thornton, and A. Sattar. Building structure into local search for SAT. In *Proceedings of the 20th International Joint Conference on Artificial Intelligence*, pages 2359–2364, 2007.

5 S. D. Prestwich. CNF encodings. In A. Biere, M. Heule, H. van Maaren, and T. Walsh, editors, *Handbook of Satisfiability* - Second Edition, volume 336 of *Frontiers in Artificial Intelligence and Applications*, pages 75–100. IOS Press, 2021.

6 P. Halmos and S. Givant. *Introduction to Boolean Algebras*. Springer, 2009.

7 G. S. Tseitin. On the complexity of derivation in propositional calculus. *Automation of Reasoning: Classical Papers in Computational Logic*, 2:466–483, 1983.

8 A. Ignatiev, A. Morgado, and J. Marques-Silva. PySAT: A Python toolkit for prototyping with SAT oracles. In O. Beyersdorff and C. Wintersteiger, editors, *Theory and Applications of Satisfiability Testing –SAT 2018. SAT 2018*, volume 10929 of Lecture Notes in Computer Science, pages 428–437. Springer, 2018.

9 A. Morgado, F. Heras, M. H. Liffiton, J. Planes, and J. Marques-Silva. Iterative and core-guided MaxSAT solving: A survey and assessment. *Constraints*, 18(4):478–534, 2013.

10 B. A. Nadel. Representation selection for constraint satisfaction: A case study using n-queens. *IEEE Expert*, 5(3):16–23, 1990.

3

SAT Encoding for AES Partial Key Recovery

After reading this chapter, you should be able to:

- Understand procedures for reducing cryptanalysis problems to SAT.
- Describe the main features of a cold boot attack and commonly adopted attack models.
- Outline the AES key expansion algorithm.
- Apply SAT to fulfill AES key recovery from a true cold boot attack.

3.1 Logical Cryptanalysis with SAT

Cryptography has been widely used in data encryption, digital signature, and message authentication. The formal analysis of the security is of utmost importance and is constantly a hot topic in the cryptographic community. SAT solvers have emerged as powerful tools in cryptanalysis, evaluating the robustness of cryptographic algorithms against potential attacks. The initial concept of applying SAT solvers to cryptanalysis was introduced by Cook and Mitchell [1]. In 1999, Massacci [2] encoded the encryption process of the Feistel block cipher Data Encryption Standard (DES) [3] into a propositional formula and analyzed it using a SAT solver. Subsequently, Massacci and Marraro [4] formally introduced logical cryptoanalysis, establishing a framework for encoding properties of crypto-algorithms into SAT problems. Since then, numerous cryptanalytic attacks leveraging SAT solvers have been developed, targeting block ciphers [5, 6], stream ciphers [7–10], asymmetric-key algorithms [11–14], and hash functions [15–17]. Logical cryptoanalysis using SAT solvers generally involves two main stages [18]:

- Encode a cryptanalysis problem into a propositional formula. Usually, bit values of plaintext, ciphertext, and the key (if used) are encoded into propositional

Applied Satisfiability: Cryptography, Scheduling, and Coalitional Games,
First Edition. Xiaojuan Liao and Miyuki Koshimura.

variables and the formula is true if and only if the ciphertext corresponds to the plaintext encrypted with the key.
- Utilize a SAT solver to identify the solution.

The challenging aspect of reducing cryptanalysis to SAT lies in finding an effective propositional encoding of a corresponding function, so that a modern SAT solver can actually crack. In practice, this task is quite complex due to the intricate nature of modern cryptographic algorithms, which are constructed from a large number of fundamental primitives. Often, a researcher has to carry out a substantial amount of manual work to create a SAT encoding for a specific cipher, either because of its unique construction features or to meet specific requirements of an implemented attack [19]. A more promising approach is to automatically transform a cryptanalysis problem to SAT by leveraging available generic systems for symbolic verification, such as CBMC [20], LLBMC [21], or domain-specific systems tailored to cryptographic areas like SAW [22], URSA [23], Grain-of-Salt [24], and TRANSALG [19]. Generic systems typically support widely employed programming languages like C, making it relatively straightforward to adapt an existing cryptographic function implementation to these tools. In contrast, domain-specific tools may be less user-friendly, but they compensate for it by providing languages purposefully enriched with specialized instructions and data types, enhancing their capability to handle cryptographic functions [19]. This richness allows for more effective treatment of cryptographic functions, although it may come at the cost of convenience compared to generic systems.

The versatility and effectiveness of SAT solving techniques have shown the potential use of SAT solvers as a tool for cryptanalysis. The difficult part at this stage is to effectively deal with parity constraints, which arise naturally in cryptanalysis, while XOR-handling in SAT solving is intractable compared to CNF-handling. For over a decade of intensive research, great attempts have been devoted to integrating XOR handling ability into SAT solving. According to various stages of transformation, there are two different strategies of handling XOR. On the one hand, noninteractive XOR-handling strategy performs XOR reasoning in the first stage of logical cryptanalysis before SAT solving. On the other hand, interactive XOR-handling invokes XOR reasoning on-the-fly during SAT solving. Positive results have been demonstrated especially by the latter strategy [9]. For instance, [25] extended the SAT solver' input language to support the XOR operation, so that the solver can handle XOR operations natively to speed up the solving of stream ciphers Crypto-1 and Grain.

Despite an impressive progress in the development of applied algorithms for solving SAT, the majority of cryptanalysis problems remain too hard even for cutting edge SAT solvers [19]. However, with the aid of side-channel attacks, modern SAT solvers display substantial potential in breaking cryptographic systems. In this chapter, we illustrate the vulnerability of a symmetric key cryptographic algorithm to a side-channel attack known as the cold boot attack. This attack enables an adversary to extract decayed keys from memory and subsequently leverage a SAT solver to reconstruct corrupted keys.

3.2 Cold Boot Attack

A Dynamic Random Access Memory (DRAM) cell is a capacitor that can be either charged or discharged, corresponding to a bit being in either a *charged* or *ground* state, respectively. DRAM remanence, introduced by Halderman et al. [26], describes the phenomenon where DRAM retains its state for several seconds after power loss, and for minutes or even hours if kept at low temperatures. The persistence of this state depends on various factors such as the hardware, temperature, and elapsed time since power removal. Cold boot attacks, detailed in [26], are sophisticated side-channel attacks that exploit DRAM remanence to extract sensitive information from a running computer. These attacks are particularly dangerous to systems that store sensitive data in memory. For example, cold boot attacks have compromised several disk encryption systems, including BitLocker, TrueCrypt, FileVault, LoopAES, and dm-crypt [27]. These disk encryption softwares often keep the encryption key in DRAM while the disk is mounted, making it vulnerable to attackers who can access DRAM contents to retrieve the key and decrypt the disk.

The cold boot attack has three variants of increasing resistance to countermeasures [27]. The most basic method involves rebooting the machine and loading a custom kernel with minimal memory usage, allowing the attacker to access residual memory. A more sophisticated approach involves briefly cutting power to the machine, then restoring it and booting a custom kernel. This method prevents the operating system from erasing memory before shutdown. The most advanced variant involves cutting the power, transferring the DRAM modules to a second prepared PC, and using that system to extract their state. This attack further deprives the original Basic Input/Output System (BIOS) and PC hardware of any chances to clear the memory on boot.

Due to the nature of cold boot attacks, memory bits start to degrade over time once power is lost, eventually leaving the attacker with only a corrupted memory image. Recovering a cryptographic key from this corrupted memory is typically done by leveraging the redundancy of key material inherent in cryptographic algorithms. In practice, many encryption programs store data precomputed from encryption keys to enhance computation speed. For block ciphers, this often involves a key schedule composed of multiple round keys pre-computed from the secret key. This precomputed data holds significantly more information than the key itself, allowing efficient reconstruction of the original key even if errors are present [27]. The advent of cold boot attacks shifted the focus of cryptologic research toward the structure of block cipher key schedules. The resistance of ciphers to such attacks has become an important measure of their security, a consideration that was previously overlooked by designers [28]. This chapter concentrates on recovering Advanced Encryption Standard (AES) keys from decayed memory bits obtained through cold boot attacks.

3.3 Advanced Encryption Standard (AES)

AES [29] is a specification for the encryption of electronic data, established by the U.S. National Institute of Standards and Technology (NIST) in 2001. This symmetric cryptographic algorithm has become one of the most widely used standards for data encryption in the world.

AES is a type of block cipher that relies on expanded keys during both the encryption and decryption processes. To enhance efficiency, the algorithm does not re-calculate the expanded key with the expansion function for each encryption or decryption operation. Instead, these expanded keys are stored in RAM until needed. This practice, however, introduces a vulnerability to cold boot attacks, where an attacker can extract the encryption key from the memory.

The AES encryption key comprises a key schedule, which includes multiple round keys generated from an *initial key* through a key expansion function. The initial key length can be 128, 192, or 256 bits, known as AES-128, AES-192, and AES-256, respectively. The key schedule of AES is a primary source of key redundancy, providing enough information for an attacker to reconstruct the initial key even if the extracted memory content has moderate errors.

Let us take AES-128 as an example to illustrate the structure of AES key schedules. An AES-128 key schedule consists of 11 round keys, each comprising 128 bits. The initial key itself serves as the 0th round key, denoted as the initial round key. This key is then mapped bijectively to the subsequent 10 round keys using the AES key expansion function. Each bit in the key schedule, either 0 or 1, is naturally expressed as a Boolean variable. The ith bit of the rth roundkey is denoted by b_i^r, where $0 \leqslant r \leqslant 10$ and $0 \leqslant i \leqslant 127$. $b_i^r = 1$ if the corresponding bit in the memory is 1 and $b_i^r = 0$ otherwise. The key schedule is addressed using two notations: the round-dependent word array and the substitution box. The round-dependent word array, represented as $R(r)$, consists of the values $\{02\}_H^{r-1}$ of the least significant byte and 0 for the remaining bytes. Here, $\{02\}_H^{r-1}$ represents the powers of $\{02\}_H$ in the Galois field $GF\left(2^8\right)$, with $\{02\}_H$ being the hexadecimal representation of 2. For each round from 1 to 10, the values of $R(r)$ $(1 \leqslant r \leqslant 10)$ are listed as follows.

$$R(1) = \underbrace{0\cdots0}_{24\text{ bits}}00000001, \quad R(6) = \underbrace{0\cdots0}_{24\text{ bits}}00100000,$$

$$R(2) = \underbrace{0\cdots0}_{24\text{ bits}}00000010, \quad R(7) = \underbrace{0\cdots0}_{24\text{ bits}}01000000,$$

$$R(3) = \underbrace{0\cdots0}_{24\text{ bits}}00000100, \quad R(8) = \underbrace{0\cdots0}_{24\text{ bits}}10000000,$$

$$R(4) = \underbrace{0\cdots0}_{24\text{ bits}}00001000, \quad R(9) = \underbrace{0\cdots0}_{24\text{ bits}}00011011,$$

$$R(5) = \underbrace{0\cdots0}_{24\text{ bits}}00010000, \quad R(10) = \underbrace{0\cdots0}_{24\text{ bits}}00110110.$$

The substitution box, abbreviated with S-box, is another crucial component used in the key expansion process. It performs a byte-wise substitution to enhance security by introducing nonlinearity into the cipher. S-box operates independently on each byte by taking the multiplicative inverse in the finite field $GF(2^8)$ using the irreducible polynomial $x^8 + x^4 + x^3 + x + 1$ and then applying an affine transformation over $GF(2)$. More details can be found in the literature [29].

In the hardware implementation of the AES key expansion function, an S-box operation could be decomposed into eight functions in algebraic normal form (ANF),[1] each with a 1-byte input and a 1-bit output [30]. Appendix B provides an example of an S-box in ANF. Each of these eight functions is represented by $S_x(B_i^r)$, where $x\ (= 0, \ldots, 7)$ is the index of the function, and B_i^r is the input byte starting with b_i^r, arranged in a least-significant-bit-first order, i.e., $B_i^r = \{b_i^r, b_{i+1}^r, \ldots, b_{i+7}^r\}$. The final output of an S-box is a byte formed by combining the outputs of these eight functions, arranged by decreasing significance of x. The formulas below describe each bit in the 1–10 round keys:

$$\begin{aligned} b_i^r &= b_i^{r-1} \oplus S_{i\,mod\,8}\left(B_{104+8\cdot\lfloor i/8\rfloor}^{r-1}\right) \oplus R_i(r), 0 \leqslant i \leqslant 23,\\ b_i^r &= b_i^{r-1} \oplus S_{i\,mod\,8}\left(B_{96}^{r-1}\right) \oplus R_i(r), 24 \leqslant i \leqslant 31,\\ b_i^r &= b_i^{r-1} \oplus b_{i-32}^r, 32 \leqslant i \leqslant 127. \end{aligned} \tag{3.1}$$

where $1 \leqslant r \leqslant 10$, $R_i(r)$ is the ith bit of the round constant word array $R(r)$, and $\lfloor x \rfloor$ is the floor function that returns the largest integer not greater than x. The relationships among the bits, as shown in Eq. (3.1), illustrate that each bit in the following 10 rounds is linked to its preceding bits. In particular, b_i^r $(0 \leqslant i \leqslant 31, 1 \leqslant r \leqslant 10)$ is determined by nine bits, i.e., b_i^{r-1} and $B_{104+8\cdot\lfloor i/8\rfloor}^{r-1}$. Meanwhile, b_i^r $(32 \leqslant i \leqslant 127, 1 \leqslant r \leqslant 10)$ is influenced by two bits, i.e., b_i^{r-1} and

1 A logical formula is in ANF if it is an XOR of a constant and one or more conjunctions of Boolean variables.

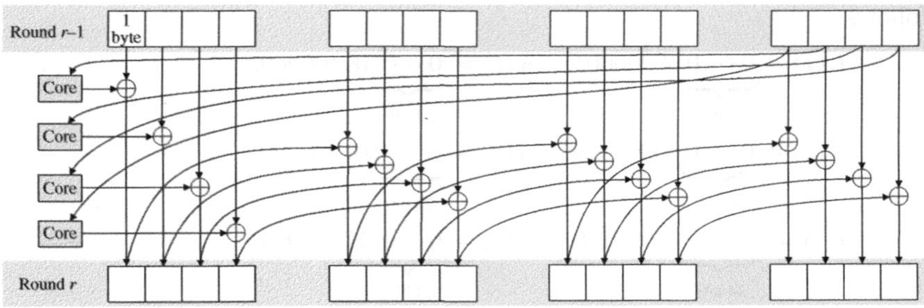

Figure 3.1 Diagram of AES-128 key expansion in adjacent two rounds (Source: Xiaojuan Liao et al., 2013/IEEE).

b^r_{i-32}. This means that an error in a bit can be corrected by examining its 2 or 9 related bits. Similarly, if multiple bits have decayed, they can be accurately recovered as long as a sufficient number of bits are known. For clarity, we refer to these relationships defined in Eq. (3.1) as *bit-relations*. The bit-relations among two adjacent rounds are depicted in Figure 3.1, where the core involves an S-box combined with the round-dependent word array $R(r)$.

3.4 Decay Pattern Assumptions and AES Key Recovery Solutions

The recovery algorithms for AES key schedules vary based on different assumptions and techniques. This section will first introduce the models related to decay patterns, followed by a survey of existing methods for recovering the AES key schedule in cold boot attack scenarios.

When memory loses power, the refresh cycle of DRAM is halted. Halderman et al. [26] observed that, over time, most memory bits tend to revert to their ground states, with a constant and small fraction of bits flipping to their charged states. Depending on how the cell is wired, the ground state can be encoded as either 0 or 1. Consequently, memory decay typically manifests as either $0 \to 1$ or $1 \to 0$. The decay direction within a given region can be inferred by comparing the numbers of 1s and 0s, as these should be approximately equal in an uncorrupted key schedule [26]. For simplicity, the remainder of this chapter assumes that 1 represents the charged state and 0 represents the ground state. Thus, the predominant decay direction is $1 \to 0$, with a small fraction of bits decaying from 0 to 1.

The probability of a 1 decaying to 0 is referred to as the *decay factor* δ_0, while the probability of a 0 flipping to 1 is denoted by δ_1. Generally, δ_0 reflects the extent of decay, increasing toward 1 as time goes on after power is lost. Conversely, δ_1 remains relatively constant and tiny, typically ranging from 0.05% to 0.1% [26].

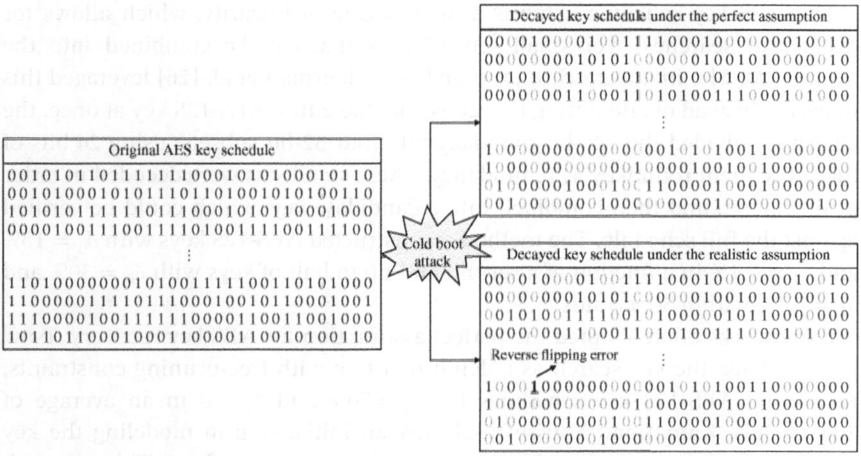

Figure 3.2 Decayed AES key schedules under different decay assumptions (Source: Xiaojuan Liao and Miyuki Koshimura/original).

Based on different values of δ_1, the research community models decay patterns using either of the following two cases.

- **Perfect assumption**: In this model, the decay direction is presumed to be exclusively from 1 to 0, given that the probability δ_1 of a 0 flipping to a 1 is negligible. Hence, δ_1 is effectively set to 0. This implies that any 1s found in the decayed key schedule are assumed to be correct without any doubt, making it easier to reconstruct the original key.
- **Realistic assumption**: This model takes into account the real-world observations where δ_1 ranged from 0.05% to 0.1%. In this scenario, both types of decay, i.e., $1 \rightarrow 0$ and $0 \rightarrow 1$, are considered, indicating that errors in both directions are possible during a cold boot attack. Consequently, none of the bits in the decayed key schedule can be deemed absolutely certain, thus adding difficulty to the key recovery process.

Figure 3.2 shows an example of decayed AES key schedules with different decay patterns. Gray bits in the figure indicate decay from 1 to 0, while the black bold underlined bit represents a reverse flipping error from 0 to 1. Under the perfect assumption, all decayed bits change from 1 to 0. In contrast, under the realistic assumption, decay occurs in both directions. The black bold underlined bit in the last round is the reverse flipping error, having flipped from 0 to 1, while all other decayed bits (highlighted in gray) turn from 1 to 0.

The technique of locating and identifying AES keys in memory was developed in [27]. The rest of this section reviews plausible techniques to reconstruct AES key schedules after they have been corrupted and extracted from memory.

The AES key schedule exhibits a high degree of linearity, which allows for exhaustive searches over small sets of keys that can be combined into the overall key. The first algorithm proposed by Halderman et al. [26] leveraged this linearity. Instead of attempting to reconstruct the entire AES-128 key at once, the algorithm divided the 128-bit roundkey into four 32-bit subsets, using 24 bits of the subsequent roundkey as redundancy. These small sets were decoded in order of likelihood, and then combined into a candidate key, which could be verified against the full schedule. The method reconstructed AES-128 keys with $\delta_0 = 15\%$ and $\delta_1 = 0.1\%$ in a fraction of a second and up to half of keys with $\delta_0 = 30\%$ and $\delta_1 = 0.1\%$ within 30 seconds.

Early works often adopted the perfect assumption for AES key recovery. Tsow [31] modeled the key search as a depth-first tree with tree-pruning constraints, recovering AES-128 key schedules for $\delta_0 = 70\%$ and $\delta_1 = 0$ in an average of 300 seconds. Kamal and Youssef [32] took an initial step in modeling the key recovery problem as a SAT problem by fully utilizing bits equal to 1. This approach recovered AES-128 key schedules with $\delta_0 = 70\%$ and $\delta_1 = 0$ in approximately 1.2 seconds on average and completed key recovery for $\delta_0 = 78\%$ and $\delta_1 = 0$ in around two hours. When δ_0 reached 80%, only 7 out of 100 instances were solved due to extended computation times.

Algorithms designed for the perfect assumption cannot recover keys if even a single bit flip occurs in the opposite direction. To address this limitation, subsequent works have employed various techniques for the realistic assumption. Based on the algorithm presented by Tsow [31] and Tanigaki and Kunihiro [33] developed a maximum likelihood approach. Theoretical analysis showed that this approach was applicable for situations where $\delta_0 < 75\%$ and $\delta_1 = 0.1\%$. Liao et al. [34] extended the work by Kamal and Youssef [32] by recasting the AES-128 key recovery problem as a partial MaxSAT problem. Using a MaxSAT solver, they recovered decayed key schedules with $\delta_0 = 76\%$ and $\delta_1 = 0.1\%$ within an average of 300 seconds. Inspired by the success of satisfiability theories applied to smaller keys, [35] combined a deep belief propagation neural network with an adapted SAT solver to achieve AES-256 key recovery under the realistic assumption. Their algorithm achieved a success rate of 96% with $\delta_0 = 65\%$ and $\delta_1 = 0.1\%$.

3.5 SAT Approach for Recovering AES Key Schedules

The superiority of SAT solvers in AES key recovery has been demonstrated by Kamal and Youssef [32], which achieved a success rate of 100% for $\delta_0 = 78\%$ under the perfect assumption. For a SAT solver, the attempt to find an assignment to variables is based on the premise that all specified constraints are satisfied without any exception, otherwise, the solver outputs unsatisfiable. This feature is well-suited for the perfect assumption, where all 1s present in the decayed AES key schedule are correct with absolute certainty. Additionally, the bit-relations in the AES key schedule can be easily formulated as a SAT problem, making it naturally suitable for SAT solvers [32]. Therefore, given a sufficient number of 1s in the corrupted key schedule, a SAT solver can infer the values of other unknown bits by formulating all the 1s and bit-relations as hard constraints.

However, in reality, reverse-flipping errors exist. Given $\delta_1 = 0.1\%$, reverse flipping is expected to occur around 1 or 2 times in a real key recovery attack for AES-128, which has 1408 total key bits, and more times for AES-192 and AES-256, which contains a larger number of bits prone to decay. Specifically, reverse flipping errors arise one to three times in AES-192 and two to four times in AES-256. Although bits flipping from 0 to 1 are very rare in number, these fatal events are sufficient to derail SAT solvers, leading them to mistake the wrong 1s as the correct constraints to infer the rest of the unknown bits. As a result, conflicts may occur during the reasoning and SAT solvers would fail to recover the correct key schedule by outputting unsatisfiable. To overcome the limitation, subsequent work [34] extended [32] to adapt SAT approach to a true cold boot attack. Below is a straightforward method to apply the SAT solver to recover AES key schedules in the presence of reverse flipping errors.

The key to addressing this problem is to identify and rectify the reverse flipping errors. A straightforward way is to conduct an exhaustive search over all 1s in a bit-by-bit fashion, exploiting the fact that a SAT solver will always output unsatisfiable as long as reverse flipping errors exist. Only if all the incorrect 1s are cleared can the solver find the correct assignment of variables. The procedure of identifying and correcting the reverse flipping errors is exhibited in Algorithm 3.1. Note that the maximum number of reverse flipping errors k is unknown to the solver, but it can be estimated as $n \cdot \delta_1$, where n is the number of 1s in the decayed key schedule. When there is one reverse flipping error, to distinguish this incorrect 1 from other correct 1s, the algorithm modifies the decayed key schedule by turning one of these 1s to 0, then inputs the modified key schedule into the SAT solver. If the solver outputs satisfiable, it indicates that the modified bit was the reverse flipping error. Otherwise, that bit is reverted. The algorithm then modifies the next 1 to 0 and runs the solver again. This process repeats until the solver outputs satisfiable, thereby locating and correcting the reverse flipping error. In the worst case, the solver needs to run $(n+1)$ times, where n is the number of 1s in the decayed key schedule. The complexity of the exhaustive search is closely related to the actual number and location of the reverse flipping errors, neither of which are known to the SAT solver. If the number of reverse flipping errors is k, in the best case, the solver needs to run $(\sum_{0 \leqslant i \leqslant k-1} C_n^i + 1)$ times, while in the worst case, it must try $\sum_{0 \leqslant i \leqslant k} C_n^i$ times, where C_n^i is the number of i-combinations of n. It is evident that a SAT solver must run multiple times to recover a decayed key schedule with reverse flipping errors. Clearly, this method is tedious and cumbersome.

Algorithm 3.1 Identify and correct reverse flipping errors in an AES key schedule.

Input: Decayed key schedule K with n bits set to 1; SAT solver S; Maximum number of reverse flipping errors k.

Output: Recovered key schedule K'.

```
id ← indices of K where K[i] = 1.
for j ← 1 to k do
    combinations ← all combinations of id taken j at a time
    for each combination c in combinations do
        K' ← K
        for each index i in c do
            K'[i] ← 0                    ▷ Flip selected bits from 1 to 0.
        end for
        if S(K') is satisfiable then
            K ← K'
            break        ▷ Found and corrected the reverse flipping errors.
        end if
    end for
    if S(K') is satisfiable then
        break                            ▷ Exit if the solution is found.
    end if
end for
return K'
```

3.6 Performance Evaluation

This section evaluates the performance of the SAT approach applied to AES-128 key recovery. All experiments were carried out on a 2.6 GHz quad-core Intel i5-2540 processor with 8GB RAM, using CryptoMiniSat [25] as the solver, which natively supports XOR operations.

The evaluation for realistic cold boot attacks was conducted with $\delta_1 = 0.1\%$. Additionally, performance under specific numbers of reverse flipping errors was investigated through the following tests:

1) Each instance contains only 1 reverse flipping error.
2) Each instance contains 2 reverse flipping errors.

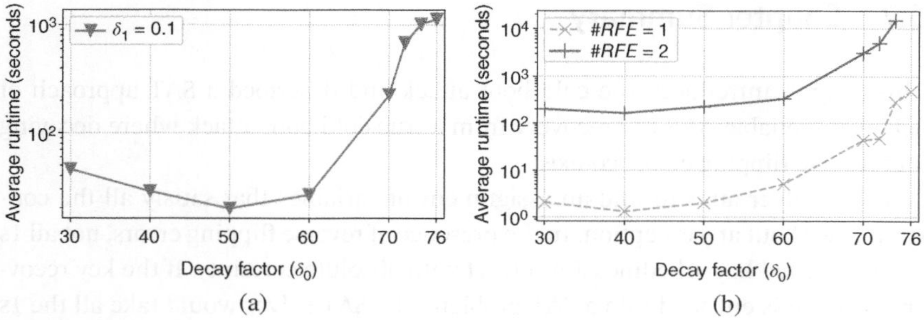

Figure 3.3 Average runtime (seconds) of SAT approach with $\delta_1 = 0.1\%$ and various numbers of Reverse Flipping Errors (RFE). (a) $\delta_1 = 0.1\%$ and (b) $\#RFE = 1$ and $\#RFE = 2$ (Source: Xiaojuan Liao and Miyuki Koshimura/original).

The results of the evaluation are summarized in Figure 3.3. It is evident that the runtime of the SAT approach under various experimental conditions shows a similar trend: it decreases and then increases as δ_0 increases. This trend is due to the combined effect of the decreasing number of times the solver needs to be run and the increasing difficulty of solving each input instance.

Initially, at a low decay factor δ_0, especially when δ_0 is 30%, the number of 1s in a decayed schedule is relatively high, leading to a significant number of runs to locate the reverse flipping errors. Although the time taken to determine the satisfiability of each input instance is quite short, the total time remains high due to the frequent runs. As δ_0 increases, the number of 1s decreases, reducing the number of solver runs required. The increase in δ_0 does not significantly impact the efficiency of CryptoMiniSat in solving each input instance, resulting in a decline in total solver time. Finally, as δ_0 rises further, the difficulty of solving each input instance increases sharply, and the time consumed by CryptoMiniSat in each run rises accordingly. Consequently, the total time spent by the SAT approach increases significantly when δ_0 exceeds 70%.

Notably, when $\#RFE = 2$, the performance of the SAT approach is consistently inferior to the case of $\#RFE = 1$ at the same decay factor δ_0. For instance, at $\delta_0 = 74\%$, the SAT solver recovered a decayed key schedule within 15 minutes when $\#RFE = 1$, whereas it took approximately four hours when $\#RFE = 2$. This demonstrates that a higher decay factor δ_0 makes it more challenging to recover a corrupted key.

In summary, the performance of the SAT approach is influenced by the number of runs needed to search for reverse flipping errors and the difficulty of solving each input instance.

3.7 Chapter Summary

This chapter introduced the cold boot attack and described a SAT approach of achieving reliable AES key recovery from a true cold boot attack where decaying and reverse flipping errors co-exist.

A SAT solver aims to find an assignment of variables that satisfy all the constraints without any exception. In the presence of reverse flipping errors, not all 1s in the decayed key schedules are correct with absolute certainty. If the key recovery problem is encoded into a SAT problem, the SAT solver would take all the 1s as hard constraints, which may result in conflicts during reasoning and eventually cause the SAT solver to fail. In order to adapt the SAT solver to a true cold boot attack, a straightforward way is to conduct repeated calls to the SAT solver until the reverse flipping errors are all corrected through bit-by-bit exhaustive search. This makes the AES key recovery possible in a true cold boot attack.

Although the conjunctive normal form (CNF) representations converted from XOR formulas could be simplified by XOR-handling enabled SAT solvers, it should be noted that the solving performance is far from satisfactory. In the next chapter, more efficiency encoding techniques will be introduced to greatly enhance the efficiency of AES key recovery from cold boot attacks.

References

1 S. A. Cook and D. G. Mitchell. Finding hard instances of the satisfiability problem: A survey. In *Satisfiability Problem: Theory and Applications, Proceedings of a DIMACS Workshop, Piscataway, New Jersey, USA, March 11–13, 1996*, volume 35 of *DIMACS Series in Discrete Mathematics and Theoretical Computer Science*, pages 1–17. DIMACS/AMS, 1996.

2 F. Massacci. Using Walk-SAT and Rel-SAT for cryptographic key search. In T. Dean, editor, *Proceedings of the 16th International Joint Conference on Artificial Intelligence, IJCAI 99, Stockholm, Sweden, July 31 - August 6, 1999. 2 Volumes*, 1450, pages 290–295. Morgan Kaufmann, 1999. URL http://ijcai.org/Proceedings/99-1/Papers/043.pdf.

3 Data Encryption Standard (DES). Federal Information Processing Standards Publication FIPS 46, 1977.

4 F. Massacci and L. Marraro. Logical cryptanalysis as a SAT problem. *Journal of Automated Reasoning*, 24(1/2):165–203, 2000. doi: 10.1023/A:1006326723002.

5 N. T. Courtois and G. V. Bard. Algebraic cryptanalysis of the data encryption standard. In *Proceedings of IMA International Conference on Cryptography and Coding*, pages 152–169, 2007.

6 N. T. Courtois, G. V. Bard, and D. Wagner. Algebraic and slide attacks on KeeLoq. In *Proceedings of International Workshop of Fast Software Encryption*, pages 97–115, 2008.

7 F. D. Garcia, G. K. Gans, R. Muijrers, P. Rossum, R. Verdult, R. W. Schreur, and B. Jacobs. Dismantling MIFARE classic. In *Proceedings of European Symposium on Research in Computer Security*, pages 97–114, 2008.

8 T. Eibach, E.Pliz, and G. Volkel. Attacking Bivium using SAT solvers. In *Proceedings of International Symposium on the Theory and Applications of Satisfiability and Testing*, pages 63–76, 2008.

9 C.-S. Han and J.-H. R. Jiang. When Boolean satisfiability meets Gaussian elimination in a simplex way. In *Computer Aided Verification - 24th International Conference, CAV 2012, Berkeley, CA, USA, July 7–13, 2012 Proceedings*, volume 7358 of *Lecture Notes in Computer Science*, pages 410–426. Springer, 2012.

10 A. A. Semenov, O. Zaikin, D. Bespalov, and M. Posypkin. Parallel logical cryptanalysis of the generator A5/1 in BNB-Grid system. In *Parallel Computing Technologies - 11th International Conference, PaCT 2011, Kazan, Russia, September 19–23, 2011. Proceedings*, volume 6873 of *Lecture Notes in Computer Science*, pages 473–483. Springer, 2011.

11 V. I. Dylkeyt, R. T. Faizullin, and I. G. Khnykin. Reducing the problem of asymmetric ciphers cryptanalysis to solving satisfiability problems. In *Proceedings of the XIII All-Russian Conference Mathematical Methods in Pattern Recognition*, pages 249–251, 2007.

12 R. T. Faizullin, I. G. Khnykin, and V. I. Dylkeyt. The SAT solving method as applied to cryptographic analysis of asymmetric ciphers. *The Computing Research Repository*, abs/0907.1755, 2009.

13 S. D. Galbraith and S. W. Gebregiyorgis. Summation polynomial algorithms for elliptic curves in characteristic two. In *Progress in Cryptology - INDOCRYPT 2014 - 15th International Conference on Cryptology in India, New Delhi, India, December 14–17, 2014, Proceedings*, volume 8885 of *Lecture Notes in Computer Science*, pages 409–427. Springer, 2014.

14 M. Trimoska, S. Ionica, and G. Dequen. Parity (XOR) reasoning for the index calculus attack. In *Principles and Practice of Constraint Programming - 26th International Conference, CP 2020, Louvain-la-Neuve, Belgium, September 7–11, 2020, Proceedings*, volume 12333 of *Lecture Notes in Computer Science*, pages 774–790. Springer, 2020.

15 I. Mironov and L. Zhang. Applications of SAT solvers to cryptanalysis of hash functions. In *Proceedings of International Symposium on the Theory and Applications of Satisfiability and Testing*, pages 102–115, 2006.

16 E. Homsirikamol, P. Morawiecki, M. Rogawski, and M. Srebrny. Security margin evaluation of SHA-3 contest finalists through SAT-based attacks. *Computer Information Systems and Industrial Management*, 7564:56–67, 2012.

17 M. Stevens, E. Bursztein, P. Karpman, A. Albertini, and Y. Markov. The first collision for full SHA-1. In J. Katz and H. Shacham, editors, *Advances in Cryptology - CRYPTO 2017 - 37th Annual International Cryptology Conference, Santa Barbara, CA, USA, August 20–24, 2017, Proceedings, Part I*, volume 10401 of *Lecture Notes in Computer Science*, pages 570–596. Springer, 2017. doi: 10.1007/978-3-319-63688-719.

18 F. Massacci. https://disi.unitn.it/~massacci/CryptoSAT/.

19 A. A. Semenov, I. V. Otpuschennikov, I. Gribanova, O. Zaikin, and S. Kochemazov. Translation of algorithmic descriptions of discrete functions to SAT with applications to cryptanalysis problems. *Logical Methods in Computer Science*, 16(1):29:1–29:42, 2020.

20 E. M. Clarke, D. Kroening, and F. Lerda. A tool for checking ANSI-C programs. In *Tools and Algorithms for the Construction and Analysis of Systems, 10th International Conference, TACAS 2004, Held as Part of the Joint European Conferences on Theory and Practice of Software, ETAPS 2004, Barcelona, Spain, March 29 –April 2, 2004, Proceedings*, volume 2988 of *Lecture Notes in Computer Science*, pages 168–176. Springer, 2004.

21 C. Sinz, F. Merz, and S. Falke. LLBMC: A bounded model checker for LLVM's intermediate representation - (competition contribution). In *Tools and Algorithms for the Construction and Analysis of Systems - 18th International Conference, TACAS 2012, Held as Part of the European Joint Conferences on Theory and Practice of Software, ETAPS 2012, Tallinn, Estonia, March 24 - April 1, 2012. Proceedings*, volume 7214 of *Lecture Notes in Computer Science*, pages 542–544. Springer, 2012.

22 K. Carter, A. Foltzer, J. Hendrix, B. Huffman, and A. Tomb. SAW: The software analysis workbench. In J. Boleng and S. T. Taft, editors, *Proceedings of the 2013 ACM SIGAda Annual Conference on High Integrity Language Technology, HILT 2013, Pittsburgh, Pennsylvania, USA, November 10–14, 2013*, pages 15–18. ACM, 2013.

23 P. Janicic. URSA: A system for uniform reduction to SAT. *Logical Methods in Computer Science*, 8(3):1–39, 2012.

24 M. Soos. Grain of salt-an automated way to test stream ciphers through SAT solvers. In *Proceedings of Workshop on Tools for Cryptanalysis*, 2010.

25 M. Soos, K. Nohl, and C. Castelluccia. Extending SAT solvers to cryptographic problems. In *Theory and Applications of Satisfiability Testing - SAT 2009, 12th International Conference, SAT 2009, Swansea, UK, June 30 - July 3, 2009. Proceedings*, volume 5584 of *Lecture Notes in Computer Science*, pages 244–257. Springer, 2009.

26 J. A. Halderman, S. D. Schoen, N. A. Heninger, W. Clarkson, W. Paul, J. A. Calandrino, A. J. Feldman, J. Appelbaum, and E. W. Felten. Lest we remember: Cold-boot attacks on encryption keys. In *Proceedings of USENIX Security Symposium*, pages 45–60, California, USA, Jul 2008.

27 N. A. Heninger. *Error Correction and the Cryptographic Key*. PhD thesis, University of Princeton, May 2011. https://www.cs.princeton.edu/techreports/2011/897.pdf.

28 M. R. Albrecht and C. Cid. Cold boot key recovery by solving polynomial systems with noise. In J. López and G. Tsudik, editors, *Applied Cryptography and Network Security - 9th International Conference, ACNS 2011, Nerja, Spain, June 7–10, 2011. Proceedings*, volume 6715 of *Lecture Notes in Computer Science*, pages 57–72, 2011.

29 Announcing the Advanced Encryption Standard (AES). Federal Information Processing Standards Publication FIPS 197, 2001.

30 X. Zhang and K. K. Parhi. High-speed VLSI architectures for the AES algorithm. *IEEE Transactions on Very Large Scale Integration (VLSI) Systems*, 12(9):957–967, 2004.

31 A. Tsow. An improved recovery algorithm for decayed AES key schedule images. In *Proceedings of Selected Areas in Cryptography*, pages 215–230, 2009.

32 A. A. Kamal and A. M. Youssef. Applications of SAT solvers to AES key recovery from decayed key schedule images. In *Proceedings of the 4th International Conference on Emerging Security Information, Systems and Technologies*, pages 216–220, 2010.

33 T. Tanigaki and N. Kunihiro. Maximum likelihood-based key recovery algorithm from decayed key schedules. In S. Kwon and A. Yun, editors, *Information Security and Cryptology - ICISC 2015 - 18th International Conference, Seoul, South Korea, November 25–27, 2015, Revised Selected Papers*, volume 9558 of *Lecture Notes in Computer Science*, pages 314–328. Springer, 2015.

34 X. Liao, H. Zhang, M. Koshimura, H. Fujita, and R. Hasegawa. Using MaxSAT to correct errors in AES key schedule images. In *25th IEEE International Conference on Tools with Artificial Intelligence, ICTAI 2013, Herndon, VA, USA, November 4–6, 2013*, pages 284–291. IEEE Computer Society, 2013.

35 I. Zimerman, E. Nachmani, and L. Wolf. Recovering AES keys with a deep cold boot attack. In M. Meila and T. Zhang, editors, *Proceedings of the 38th International Conference on Machine Learning, ICML 2021, 18–24 July 2021, Virtual Event*, volume 139 of *Proceedings of Machine Learning Research*, pages 12955–12966. PMLR, 2021.

4

MaxSAT Encoding for AES Partial Key Recovery

After reading this chapter, you should be able to:

- Apply MaxSAT to fulfill AES key recovery from a true cold boot attack.
- Discuss the pros and cons of SAT and MaxSAT approaches applied to AES key recovery.
- Understand ways to improve the AES key recovery efficiency with satisfiability theories.

4.1 Original Partial MaxSAT Approach

In Chapter 3, we have demonstrated the feasibility of utilizing a SAT solver to achieve Advanced Encryption Standard (AES) [1] key recovery in a genuine cold boot attack. However, when reverse flipping errors are present, one has to run the SAT solver repeatedly to detect and correct the reverse flipping errors. Consequently, the performance of the SAT-based approach is inevitably impacted by the frequency of solver calls. In light of this situation, it is natural to contemplate whether there exists a more favorable method to circumvent the need for multiple solver invocations.

The answer is affirmative. Drawing inspiration from MaxSAT's capability to satisfy as many constraints as possible while eliminating the minority of unsatisfied ones, Liao et al. [2] proposed reformulating the AES key recovery problem under realistic conditions as a partial MaxSAT problem. This method ensures that a single solver call is sufficient for key recovery. This section introduces the original partial MaxSAT approach from [2], with an advanced version to be discussed in Section 4.2.

The key recovery problem can be more effectively resolved using a partial MaxSAT solver. This is expected because the majority of 1s are correct, with

Applied Satisfiability: Cryptography, Scheduling, and Coalitional Games, First Edition. Xiaojuan Liao and Miyuki Koshimura.

only a small fraction being erroneous flips from 0. The goal is to identify these reverse flipping errors and recover the true bits. A partial MaxSAT solver treats bit-relations as hard constraints and considers bits equal to 1 as soft constraints. This means it accounts for the possibility that some 1s may be incorrect. Solving the partial MaxSAT problem involves finding an assignment of variables that satisfy all hard constraints and the maximum number of soft constraints. As long as the reverse flipping errors are a small percentage of all 1s, a partial MaxSAT solver can effectively clear them by satisfying the majority of soft constraints and excluding the unsatisfied minority. This approach eliminates the need for exhaustive searches over all 1s.

Modeling 1s as soft clauses is straightforward. Specifically, if the ith bit of the rth round key is 1 in the decayed key schedule, declaring $b_i^r = 1$ as a soft clause for a MaxSAT solver is sufficient. In contrast, modeling bit-relations as hard clauses is more complex, mainly due to the need to handle the XOR operation, which is crucial for characterizing bit-relations. Since MaxSAT solvers do not natively support the XOR operation, converting formulas with XOR to CNF is challenging. The remainder of this subsection will detail how to describe two types of formulas in CNF representations.

1) **XOR formula**: A formula connected by XOR operator, where all the terms are individual Boolean variables.
2) **ANF formula**: A formula connected by XOR operator, where terms consist of a constant (0 or 1) and conjunctions of Boolean variables.

A straightforward method for representing an XOR formula in CNF introduces 2^{n-1} clauses, where n is the size of the XOR formula. This approach, known as direct conversion, is suitable when the XOR formula is of relatively small size.

Example 4.1 In an AES-128 key schedule, to describe the bit-relation $b_i^r = b_i^{r-1} \oplus b_{i-32}^{r-1}$ $(32 \leqslant i \leqslant 127, 1 \leqslant r \leqslant 10)$ in CNF, only the following four clauses are sufficient:

$$\begin{aligned} &\neg b_i^r \vee b_i^{r-1} \vee b_{i-32}^{r-1}, \qquad b_i^r \vee \neg b_i^{r-1} \vee b_{i-32}^{r-1},\\ &b_i^r \vee b_i^{r-1} \vee \neg b_{i-32}^{r-1}, \qquad \neg b_i^r \vee \neg b_i^{r-1} \vee \neg b_{i-32}^{r-1}. \end{aligned}$$

For large n, apparently, this interpretation is unmanageable as the number of clauses increases exponentially with the size of an XOR formula. To address this exponential explosion, a long XOR formula is typically divided into manageable groups, each represented by an auxiliary Boolean variable. The size of those groups is called *cutting number*, denoted by k. For each group of k Boolean variables, an auxiliary variable is introduced, and the number of clauses to describe one group is no more than 2^k. The long XOR formula is then represented as a new

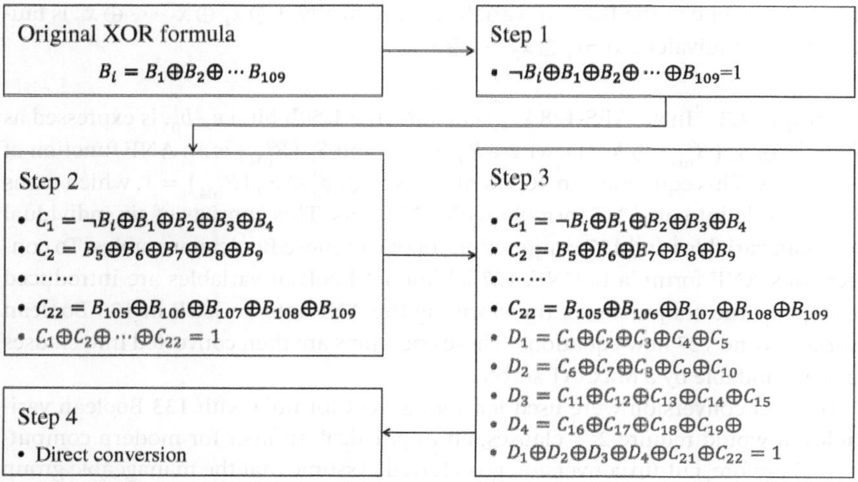

Figure 4.1 An example of applying the cut-up conversion (Source: Xiaojuan Liao and Miyuki Koshimura/original).

XOR of $\lceil n/k \rceil$ auxiliary variables, along with a set of clauses for the manageable groups. If the size of the new XOR formula is still overlong, more auxiliary Boolean variables are added to further divide the formula. This process continues until the size of the new XOR formula is manageable. This method is called as cut-up conversion. Figure 4.1 illustrates the main steps involved in applying the cut-up conversion to a long XOR formula, with the cutting number set to 5. Each step produces formulas equivalent to those in the previous step. In this example, auxiliary Boolean variables $C_1, \ldots, C_{22}$ and $D_1, \ldots D_4$ are generated to cut up the long formula $\neg B_l \oplus B_1 \oplus \cdots \oplus B_{109} = 1$ and $C_l \oplus C_1 \oplus B_2 \cdots \oplus C_{22} = 1$, respectively. Specifically, gray formulas (the last five) in Step 3 highlight the cutting up result of the gray formula (the last one) in Step 2.

So far, we have discussed methods for representing both short and long XOR formulas in CNF. Next, we will cover how to encode algebraic normal form (ANF) into conjunctive normal form (CNF). First, additional Boolean variables are introduced to replace conjunctions, transforming the ANF formula into an XOR, which can be handled by either direct or cut-up conversion. Formally, a conjunction of Boolean variables, denoted by $x_0 \wedge x_1 \wedge \cdots \wedge x_n$, could be represented by a new Boolean variable a with $n+1$ additional clauses:

$$x_1 \vee \neg a,\ x_2 \vee \neg a, \ldots, x_n \vee \neg a, \neg x_1 \vee \neg x_2 \cdots \vee \neg x_n \vee a.$$

Thus, the ANF formula is converted to a set of clauses and an XOR of a constant and several additional Boolean variables. If the constant is 0, it can be safely removed without further conversion. Otherwise, the constant 1 can be eliminated by negating one of the Boolean variables. Specifically, $1 \oplus x_1 \oplus x_2 \cdots \oplus x_n$ is tautologically equivalent to $\neg x_1 \oplus x_2 \cdots \oplus x_n$.

Example 4.2 In an AES-128 key schedule, the 129th bit, i.e., b_0^1, is expressed as $b_0^1 = b_0^0 \oplus S_0\left(B_{104}^0\right) \oplus R_0(1)$, where $R_0(1) = 1$ and $S_0\left(B_{104}^0\right)$ is an ANF function of 131 terms. This equation can be rewritten as $b_0^1 \oplus b_0^0 \oplus S_0\left(B_{104}^0\right) = 1$, which turns the left side into an ANF formula with 133 terms. This consists of six individual Boolean variables and 127 conjunctions of two or more Boolean variables. To convert this ANF formula to CNF, 127 additional Boolean variables are introduced to replace the conjunctions, transforming the ANF into an XOR of 133 Boolean variables and 127 new equations. These equations are then converted into clauses understandable by a MaxSAT solver.

If direct conversion were used for a long XOR formula with 133 Boolean variables, it would require 2^{132} clauses, an impractical number for modern computers. Therefore, cut-up conversion is preferred. Assume that the manageable group size, called the *cutting number* k, is set to 5. Then the 133 Boolean variables are divided into $\lceil 133/5 \rceil = 27$ groups, each represented by an auxiliary Boolean variable. This step introduces $2^5 \cdot 26 + 2^3 = 840$ clauses. The long XOR formula is then reduced to one involving 27 auxiliary variables and 840 clauses. However, this new XOR formula is still too long to be converted into a manageable CNF, as it would require 2^{26} clauses. To further simplify, the formula is divided again, introducing $\lceil 27/5 \rceil = 6$ auxiliary variables and $2^5 \cdot 5 + 2^2 = 164$ clauses. Finally, this reduced formula with 6 variables can be directly converted into 32 clauses. In summary, to describe the long XOR formula in CNF, a total of $27 + 6 = 33$ auxiliary variables and $840 + 164 + 32 = 1036$ clauses are needed.

Although the cut-up conversion for long XOR formulas is far from optimal, it significantly reduces the number of clauses needed to describe the S-box, which is represented by eight ANF functions containing over 100 conjunctions of Boolean variables. By introducing auxiliary Boolean variables. By using auxiliary Boolean variables to break down long XOR formulas, the total clause count for the S-box is greatly diminished.

4.2 Improved Partial MaxSAT Approach

The AES-128 key schedule includes 11 rounds, with the initial round (the 0th round) being the 128-bit initial key, and the expanded 10-round key bits derivable from this initial key. Consequently, the main objective is to recover the 128-bit initial key, rather than the entire 1408-bit key schedule. With this focus, [3] streamlined the key recovery problem by eliminating certain variables and formulas from the expanded rounds.

Definition 4.1 A variable b_i^r $(r > 0)$ in the expanded rounds is termed a leaf variable if it appears in only one formula. The formula associated exclusively with the leaf variable b_i^r is referred to as a leaf formula, denoted by f_i^r. The pair (b_i^r, f_i^r) is identified as a *leaf set*.

In a complete AES-128 key schedule, there are 32 leaf variables, i.e., the last 32 bits in the last round, denoted as b_i^{10} where $i \in \{96, \dots, 127\}$. The corresponding leaf formula f_i^{10} for each leaf variable is $b_i^{10} = b_i^9 \oplus b_{i-32}^{10}$. By contrast, any other variable in the expanded rounds appears in more than one formula: one formula describes how the variable is computed from preceding variables, while the others detail its contribution to subsequent variables. Clearly, leaf variables have weaker dependencies on other variables compared to nonleaf variables. This is because each leaf variable b_i^{10} is contained by only one formula f_i^{10}. Without f_i^{10}, the value of b_i^{10} becomes unconstrained and could potentially be omitted in key expansion.

On the other hand, in the MaxSAT approach, variables set to 1 are highly likely (around 99.9%) to be correct and treated as soft clauses by the MaxSAT solver. Conversely, variables set to 0 contribute minimal information to the problem solving because a 0 could either be the original value or a result of decay, with the decay factor δ_0 being completely unknown to the solver. Hence, variables set to 0 are considered less informative compared to those set to 1.

The improved approach involves identifying and removing leaf sets, i.e., the combinations of leaf variable and the corresponding leaf formula, where all variables in the leaf formula are set to 0. Since leaf variables have relatively weak connections to the entire key schedule, and bits equal to 0 provide little useful information to the solver, removing such leaf sets is unlikely to result in significant loss of crucial information. Instead, this elimination reduces the scale of the key recovery problem, thereby enhancing problem-solving efficiency.

Algorithm 4.1 illustrates the process of simplifying AES-128 key recovery by identifying and removing leaf sets from the final round. The input $\{B, F, S\}$ represents a complete AES-128 key recovery problem, where B and F consist of 1408 bits and corresponding formulas, and S denotes variables set to 1 in the corrupted key schedule. The algorithm outputs a simplified key recovery problem $\{B', F', S\}$, where B' and F' are subsets of B and F, respectively. This algorithm efficiently identifies and removes leaf sets from the AES-128 key recovery problem, focusing on variables in the last round that have limited impact on the overall solution. Leaf sets are checked iteratively to determine if all variables in their associated formula

Algorithm 4.1 Simplify the AES key recovery problem by removing leaf sets in the last round.

Input: A complete AES-128 key recovery problem $\{B, F, S\}$; $B = \{b_i^r\}$ is a set of key bits, $i \in \{0, \dots, 127\}$, $r \in \{0 \dots, 10\}$; $F = \{f_i^r\}$ is a set of formulas characterizing bit-relations, $i \in \{0, \dots, 127\}$, $r \in \{0 \cdots, 10\}$; S is a set of variables that present 1 in the corrupted key schedule.

Output: A simplified AES-128 key recovery problem $\{B', F', S\}$.

$LB \leftarrow \varnothing$; $LF \leftarrow \varnothing$; $RB \leftarrow \varnothing$; $RF \leftarrow \varnothing$. ▷ LB and LF is a set of leaf variables and formulas, respectively; RB and RF is a set of removable variables and formulas, respectively.
for $i = 96, 97, \dots, 127$ **do**
    $LB.push\left(b_i^{10}\right)$; $LF.push\left(f_i^{10}\right)$;
end for
while $LB \neq \varnothing$ **do**
    $b_i^r \leftarrow LB.pop$; $f_i^r \leftarrow LF.pop$;
    if S contains none of variables in f_i^r **then**
        $RB.push\left(b_i^r\right)$; $RF.push\left(f_i^r\right)$; ▷ $\left(b_i^r, f_i^r\right)$ could be removed.
        if $i > 31$ **then**
            LB.push(b_{i-32}^r); LF.push(f_{i-32}^r); ▷ Since b_i^r is removable, b_{i-32}^r becomes the new leaf bit, the successor of b_i^r.
        end if
    end if
end while
$B' \leftarrow B \setminus RB$; $F' \leftarrow F \setminus RF$;
return $\{B', F', S\}$;

are 0, indicating removable constraints that can safely be omitted from the key recovery process.

To facilitate understanding, Figure 4.2 illustrates the constraints applied to bits b_{127}^{10}, b_{95}^{10}, b_{63}^{10}, and b_{31}^{10}. We start the analysis from the last bit of the last block, denoted by b_{127}^{10}. According to the structure of AES, there is only one formula related to b_{127}^{10}:

$$b_{127}^{10} = b_{127}^{9} \oplus b_{95}^{10} \tag{4.1}$$

As stated in Algorithm 4.1, if $b_{127}^{10} = b_{127}^{9} = b_{95}^{10} = 0$ in the decayed key schedule, it means that the bit b_{127}^{10} contributes nothing to the solver to infer the values of other variables. Therefore, b_{127}^{10} and Formula (4.1) can be removed safely. All the bits in the last block (dotted filled block in Figure 4.2) are generally independent of each other, and can be dealt with in the same way. After dealing with the last block by removing some bits, we move to handle the last second block of the last round (diagonal line filled block in Figure 4.2). If b_{127}^{10} and Formula (4.1) has been

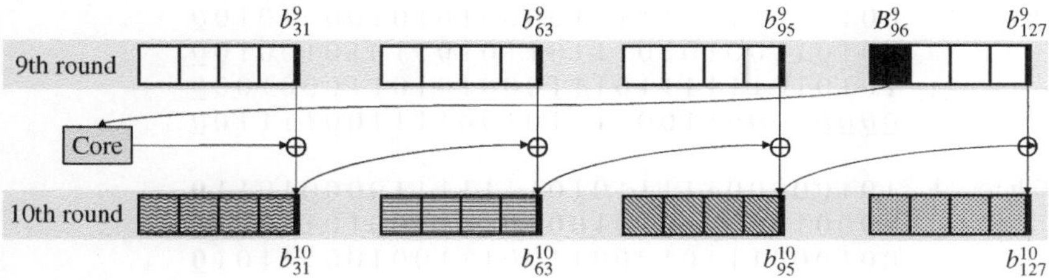

Figure 4.2 Bit relations related to b^{10}_{127}, b^{10}_{95}, b^{10}_{63}, and b^{10}_{31}. B^9_{96} is the input byte starting with b^9_{96}, arranged in a least-significant-bit-first order, i.e., $B^9_{96} = \{b^9_{96}, b^9_{97}, b^9_{98}, b^9_{99}, b^9_{100}, b^9_{101}, b^9_{102}, b^9_{103}\}$ (Source: Xiaojuan Liao and Miyuki Koshimura/original).

removed, then b^{10}_{95} becomes the last bit. We move on considering b^{10}_{95}. Besides Formula (4.1) that has been deleted, there is one more formula related to b^{10}_{95}:

$$b^{10}_{95} = b^9_{95} \oplus b^{10}_{63} \tag{4.2}$$

If all the bits shown in Formula (4.2) are 0, then b^{10}_{95} and Formula (4.2) can be removed. Similar method could be applied to the rest bits of the last second block, as well as the last third block (horizontal line filled block in Figure 4.2).

For the first block of the last round, the handling method is somewhat different because in this block, S-box function is involved. If b^{10}_{127}, b^{10}_{95}, and b^{10}_{63} and all the formulas related to these three variables have been deleted, then b^{10}_{31} becomes the last bit of the whole key schedule.[1] Then we consider the value of b^{10}_{31} and the formulas related to b^{10}_{31}:

$$b^{10}_{31} = b^9_{31} \oplus S_0\left(\left\{b^9_{96}, b^9_{97}, b^9_{98}, b^9_{99}, b^9_{100}, b^9_{101}, b^9_{102}, b^9_{103}\right\}\right) \oplus R_{31}(10) \tag{4.3}$$

where $R_{31}(10)$ is a one-bit constant. If all the bits shown in Formula (4.3) are 0, then b^{10}_{31} and Formula (4.3) can be removed. All 32 bits in the first block of the last round (wavy line filled block in Figure 4.2) can be dealt with in the same way. Figure 4.3 shows the last two adjacent rounds of a decayed AES key schedule, in which b^{10}_{127}, b^{10}_{95}, b^{10}_{63}, and b^{10}_{31} are all leaf variables (bold and underlined). Variables involved in leaf formulas are highlighted in italics and underlined. The leaf variables and their associated formulas can be all removed.

It is noteworthy that while leaf sets are primarily targeted in the final round, similar elimination may extend to earlier rounds ($(10-m)$th round for $1 \le m \le 9$). However, the feasibility of removal diminishes with earlier rounds due to intricate dependencies from multiple formulas involving each variable. For instance, variables in the last block of the 9th round are involved in eight S-box functions and two short XOR functions. Removing such variables requires

1 In fact, some bits in the horizontal line-filled, diagonal line-filled, and dotted filled blocks have not been removed., but their values are not restricted by b^{10}_{31}.

9th round	0 0 1 0 1 0 1 1 0 1 1 1 1 1 0 0 0 0 1 0 1 0 1 0 0 0 1 0 1 0 0 *0*
	0 0 1 0 1 0 0 0 1 0 1 0 1 1 1 0 1 1 0 1 0 0 1 0 1 0 1 0 0 1 0 *0*
	1 0 1 0 1 0 1 1 1 1 1 1 0 1 1 1 0 0 0 1 0 1 0 1 1 0 0 0 1 0 0 *0*
	0 0 0 0 0 0 0 0 1 1 0 0 1 1 1 1 0 1 0 0 1 1 1 1 0 0 1 1 1 1 0 *0*
10th round	1 0 1 0 0 0 0 0 1 1 1 1 1 0 1 0 1 1 1 1 1 1 1 0 0 0 0 1 0 1 1 **0**
	1 0 0 0 1 0 0 0 0 1 0 1 0 1 0 0 0 0 1 0 1 1 0 0 1 0 1 1 0 0 1 **0**
	0 0 1 0 0 0 1 1 1 0 1 0 0 0 1 1 0 0 1 1 1 0 0 1 0 0 1 1 1 0 1 **0**
	0 0 1 0 1 0 1 0 0 1 1 0 1 1 0 0 0 1 1 1 0 1 1 0 0 0 0 0 0 1 1 **0**

Figure 4.3 An example in which b^{10}_{127}, b^{10}_{95}, b^{10}_{63}, and b^{10}_{31} are all leaf variables (Source: Xiaojuan Liao and Miyuki Koshimura/original).

ensuring all corresponding variables in these 10 functions are 0, a stringent condition that is difficult to satisfy.

4.3 Performance Evaluation

This section shows a performance comparison between Boolean satisfiability (SAT) approach (which was introduced in Chapter 3) and two partial MaxSAT approaches applied to AES-128 key recovery. The SAT approach is evaluated using CryptoMiniSat [4], a solver that natively supports XOR operations. For the partial MaxSAT approaches, Pwbo2.0 was chosen as the solver, demonstrating superior performance compared to other contemporary candidates [2]. In the experiments, the size k of groups in the cut-up conversion is set to 5, a configuration shown to be more effective as the decay factor δ_0 increases [3].

4.3.1 Results of SAT and Original Partial MaxSAT Approaches

This subsection evaluates the SAT and Original partial MaxSAT Approach (OMA) under various numbers of reverse flipping errors. Both approaches were tested on a 2.6 GHz quad-core Intel i5-2540 processor with 8GB RAM.

The evaluation results for true cold boot attacks (where $\delta_1 = 0.1\%$) are summarized in Table 4.1. It is evident from the data that OMA consistently outperforms the SAT approach in terms of efficiency across various decay factors. Unlike the SAT approach, where performance is influenced by both the number of solver runs and the complexity of solving individual instances in each run, the runtime of OMA shows a direct correlation with increasing δ_0. This indicates that the efficiency of OMA is predominantly determined by the difficulty of achieving key recovery at a specific δ_0.

Table 4.1 Average runtime (seconds) of SAT approach and original partial MaxSAT approach (OMA) at $\delta_1 = 0.1\%$.

δ_0 (%)	30	40	50	60	70	72	74	76
SAT	45.81	28.47	19.66	26.52	225.38	678.45	1004.16	1116.35
OMA	0.94	0.96	1.17	1.56	12.53	26.78	231.61	296.42

To explore the performances of SAT and MaxSAT approaches under specific number reverse flipping errors, additional test were conducted across two scenarios:

1) Each instance contains only 1 reverse flipping error.
2) Each instance contains 2 reverse flipping errors.

Table 4.2 presents the average runtime comparison between SAT and OMA for these cases. It is evident that OMA exhibits slightly faster runtime than the SAT approach when dealing with instances containing a single reverse flipping error in the decayed key schedule. Moreover, OMA demonstrates significantly superior performance compared to SAT when the instances involve exactly two reverse flipping errors, across all decay factors. The reduced efficiency of the SAT approach can be attributed to the extensive searching required for identifying reverse flipping errors, whereas the MaxSAT solver executes a single optimized run to exclude these minimal errors.

Table 4.2 Average runtime (seconds) of SAT approach and original partial MaxSAT approach (OMA) with 1 and 2 reverse flipping errors (*#RFE*).

Decay factor	*#RFE* = 1		*#RFE* = 2	
δ_0 (%)	SAT	OMA	SAT	OMA
30	2.05	1.04	198.64	1.46
40	1.31	1.09	162.25	1.56
50	1.97	1.35	224.69	2.18
60	5.03	2.25	329.62	4.68
70	43.53	14.95	3047.82	47.73
72	47.60	28.35	4909.57	245.18
74	280.83	122.52	14715.61	2160.49
76	480.10	384.35	—	—

4.3.2 Results of Two Partial MaxSAT Approaches

This subsection presents a comparative analysis between the Original partial MaxSAT Approach (OMA) and the Improved partial MaxSAT Approach (IMA) Both approaches were tested on a 2.0 GHz quad-core Intel i7-4510U processor with 8GB RAM.

Table 4.3 summarizes the performance results under varying decay factors (δ_0). For smaller decay factors, specifically when δ_0 is up to 60%, both approaches demonstrate similar performance levels. However, as δ_0 exceeds 60%, IMA exhibits greater time efficiency compared to OMA. Notably, at a decay factor of 76%, OMA requires approximately 358 seconds on average to recover corrupted key schedules with 1 reverse flipping error, whereas IMA accomplishes the same task in just 300 seconds.

Figure 4.4 illustrates the average numbers of variables and clauses in both OMA and IMA, shedding light on why IMA outperforms OMA. Notably, regardless of whether there is one or two reverse flipping errors in solving the same instances, IMA consistently exhibits fewer variables and clauses compared to OMA. In general, when the problem complexity remains unchanged, reducing the number of variables and clauses in MaxSAT computations correlates with reduced computation time. This reduction in computational elements explains why IMA achieves greater time efficiency than OMA at a fixed δ_0. However, it's important to note that despite the decrease in the number of variables and clauses as δ_0 increases, the average runtime increases significantly. This phenomenon occurs because higher decay factors intensify the difficulty of the key recovery problem, thereby impacting the efficiency of the problem-solving process.

Table 4.3 Average runtime (seconds) of original partial MaxSAT approach (OMA) and improved partial MaxSAT approach (IMA) with 1 and 2 reverse flipping errors (*#RFE*).

Decay factor	***#RFE* = 1**		***#RFE* = 2**	
δ_0 (%)	**OMA**	**IMA**	**OMA**	**IMA**
30	0.78	0.77	1.02	1.00
40	0.82	0.85	1.16	1.17
50	0.99	1.03	1.61	1.62
60	1.59	1.64	4.28	3.92
70	10.84	10.46	38.15	36.04
72	17.44	16.35	151.46	133.19
74	246.09	192.93	1802.97	1638.23
76	358.79	301.60	—	—

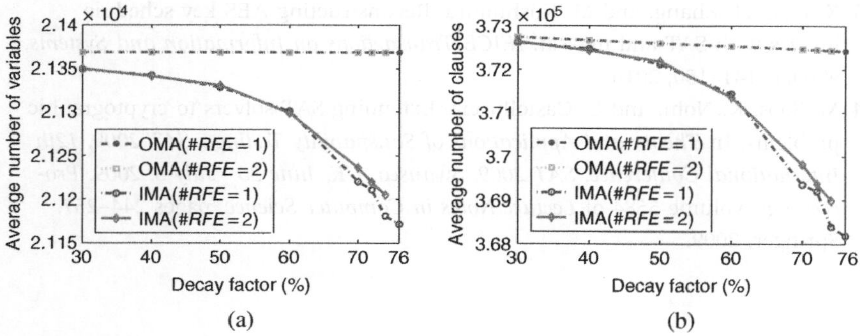

Figure 4.4 Average number of variables and clauses generated by two partial MaxSAT approaches. (a) Number of variables (b) number of clauses (Source: Xiaojuan Liao et al. (2016)/with permission of IEICE Transactions on Information and Systems).

4.4 Chapter Summary

This chapter introduced two partial MaxSAT approaches of achieving reliable AES key recovery from a true cold boot attack. The key recovery problem is formulated as a partial MaxSAT problem, where bit-relations are treated as hard constraints and all 1s as soft constraints. A partial MaxSAT solver aims to satisfy as many soft constraints as possible while minimizing the number of unsatisfied ones, thereby solving the key recovery problem in a single run.

One drawback of the MaxSAT approach is the absence of native support for XOR operations in solvers. To handle XOR constraints, a large number of variables and clauses must be introduced for a static XOR-CNF conversion. Although the XOR-handling strategy in MaxSAT may not be as efficient as in SAT, the overall performance of the MaxSAT approach is still significantly better than that of the SAT approach. Moreover, by leveraging the specific structure of AES key schedules, the efficiency of the partial MaxSAT approach can be further optimized.

References

1 Announcing the Advanced Encryption Standard (AES). Federal Information Processing Standards Publication FIPS 197, 2001.

2 X. Liao, H. Zhang, M. Koshimura, H. Fujita, and R. Hasegawa. Using MaxSAT to correct errors in AES key schedule images. In *25th IEEE International Conference on Tools with Artificial Intelligence, ICTAI 2013, Herndon, VA, USA, November 4–6, 2013*, pages 284–291. IEEE Computer Society, 2013.

3 X. Liao, H. Zhang, and M. Koshimura. Reconstructing AES key schedule images with SAT and maxsat. *IEICE Transactions on Information and Systems*, 99-D(1):141–150, 2016.

4 M. Soos, K. Nohl, and C. Castelluccia. Extending SAT solvers to cryptographic problems. In *Theory and Applications of Satisfiability Testing - SAT 2009, 12th International Conference, SAT 2009, Swansea, UK, June 30 - July 3, 2009. Proceedings*, volume 5584 of *Lecture Notes in Computer Science*, pages 244–257. Springer, 2009.

5

Job-Shop Scheduling

After reading this chapter, you should be able to:

- Understand the model of Job-Shop Scheduling Problem (JSSP).
- Comprehend the process of encoding a JSSP into a Boolean Satisfiability (SAT) problem.
- Explain the phenomenon observed in experiments: why a SAT solver requires more time to solve the smaller-scale problems.

5.1 Job-shop Scheduling Model

Scheduling is a decision-making process that is used on a regular basis in many manufacturing and services industries. It deals with the allocation of resources to tasks over given time periods, and its goal is to optimize one or more objectives [1]. Algorithms for solving scheduling are typically categorized into two main groups: approximation methods and exact methods. In cases where scheduling problems are large in scale and computationally intractable, exact methods may require significantly prolonged computation times to find optimal solutions. As an alternative, approximation methods are proposed to efficiently seek suboptimal solutions within a shorter computation time [2–4]. Nevertheless, it remains highly significant to develop optimization algorithms, even as a means to evaluate suboptimal solutions. Such algorithms can yield substantial benefits, especially when the scheduled application needs to be executed multiple times [5].

Efforts to find optimal solutions for scheduling problems have traditionally centered around dynamic programming and branch-and-bound algorithms [6, 7]. In the past decade, there has been a notable shift toward formalizing scheduling problems using generalized models, such as mathematical programming [8–11],

Applied Satisfiability: Cryptography, Scheduling, and Coalitional Games,
First Edition. Xiaojuan Liao and Miyuki Koshimura.

satisfiability modulo theory [5, 12–14], Boolean Satisfiability (SAT) [15–18], and partial MaxSAT [19–23]. This shift has garnered significant attention. Motivated by the tremendous progress made in academically and commercially available software designed for addressing these generalized problems, scheduling problems can now be efficiently solved in an out-of-the-box manner, requiring no specialized knowledge in scheduling or coding experience from the user [9]. For a comprehensive understanding of deterministic scheduling models and related combinatorial problems, you may refer to [1].

As a pioneering work that significantly influenced the scheduling community, Crawford and Baker [15] encoded scheduling problems into propositional logical formulas. This groundbreaking approach paved the way for subsequent research [16], which successfully addressed six types of open Job-Shop Scheduling Problems (JSSPs). This chapter demonstrates the SAT encoding method applied to six open JSSPs, i.e., ABZ8, ABZ9 [24], YN1, YN2, YN3, and YN4 [25]. It is worth noting that ABZ8 and ABZ9 are among the 10 renowned tough problems in this field.

A JSSP is a combinatorial optimization problem, which consists of a set of n jobs $\{J_1, \ldots, J_n\}$ and a set of m machines $\{M_1, \ldots, M_m\}$. Each job J_l is a sequence of operations $\langle O_1^l, \ldots, O_{q_l}^l \rangle$. Each operation O_i^l requires the exclusive use of a machine $M_{O_i^l}(1 \leq O_i^l \leq m)$ for an uninterrupted duration p_i^l, its processing time. A schedule is a set of start times for each operation O_i^l. The time required to complete all the jobs is called the makespan L. The objective of the JSSP is to determine the schedule which minimizes L.

Example 5.1 Consider a set of three jobs $\{J_1, J_2, J_3\}$ to be processed on three machines $\{M_1, M_2, M_3\}$. J_1 consists of two operations, while J_2 and J_3 consist of three operations each, i.e., $J_1 = \langle O_1^1, O_1^2 \rangle$, $J_2 = \langle O_2^1, O_2^2, O_2^3 \rangle$, $J_3 = \langle O_3^1, O_3^2, O_3^3 \rangle$. The processing information for each operation is shown in Table 5.1, with each item in parentheses indicating the processing time of the corresponding operation allocated on its available machine. Figure 5.1 illustrates two feasible solutions to this problem. Clearly, the allocation in Figure 5.1b outperforms that in Figure 5.1a, as it achieves shorter makespan.

Table 5.1 Processing information for each operation.

Job	Operation 1	Operation 2	Operation 3
J_1	$(M_1, 3)$	$(M_2, 4)$	—
J_2	$(M_2, 2)$	$(M_1, 3)$	$(M_3, 3)$
J_3	$(M_1, 4)$	$(M_3, 6)$	$(M_2, 2)$

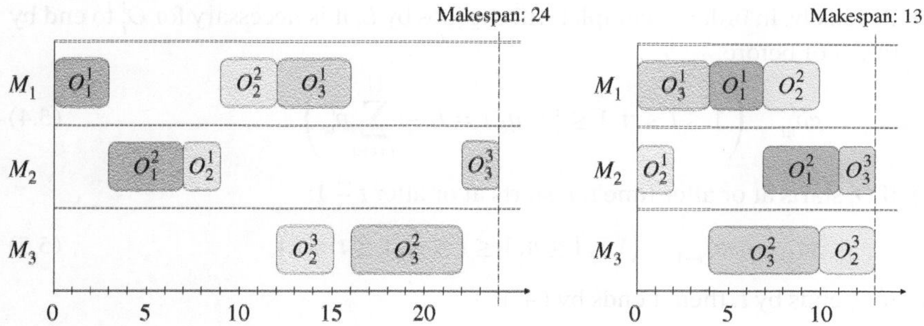

(a) (b)

Figure 5.1 Two feasible schedules in Example 5.1. (a) Feasible schedule 1 and (b) feasible schedule 2 (Source: Xiaojuan Liao and Miyuki Koshimura/original).

5.2 SAT Approach

This section outlines the process of encoding the JSSP as a SAT problem. A SAT instance comprises several clauses that a SAT solver must handle. To capture all the essential constraints defining the scheduling model, we introduce three types of propositional variables.

- $pr_{i,j}^{l,k}$ means that O_i^l precedes O_j^k.
- $sa_{i,t}^l$ means that O_i^l starts at time t or later.
- $eb_{i,t}^l$ means that O_i^l ends by time t or before.

The JSSP can be translated into a SAT problem by the following rules. In this translation, it is assumed that the makespan is at most L.

1) O_i^l precedes O_{i+1}^l:

$$p_{i,i+1}^{l,l} \quad (1 \le l \le n, 1 \le i < q_l) \tag{5.1}$$

2) If O_i^l and O_j^k require the same machines, then either O_i^l precedes O_j^k or O_j^k precedes O_i^l. That is, if $M_{O_i^l} = M_{O_j^k}$, then we add the clause:

$$p_{i,j}^{l,k} \vee p_{j,i}^{k,l} \quad (1 \le l < k \le n, 1 \le i \le q_l, 1 \le j \le q_k) \tag{5.2}$$

3) O_i^l is not able to start before all previous operations $O_1^l, \ldots,$ and O_{i-1}^l end. It requires at least $t = \sum_{u=1}^{i-1} p_u^l$. That is, O_i^l starts at time t or later:

$$sa_{i,t}^l \quad \left(1 \le l \le n, 1 \le i \le q_l, t = \sum_{u=1}^{i-1} p_u^l\right) \tag{5.3}$$

4) Reversely, in order to complete all the jobs by L, it is necessary for O_i^l to end by time t or before:

$$eb_{i,t}^l \quad \left(1 \le l \le n, 1 \le i \le q_l, t = L - \sum_{u=i+1}^{q_l} p_u^l\right) \tag{5.4}$$

5) If O_i^l starts at or after time t, it starts at or after $t-1$:

$$sa_{i,t}^l \rightarrow sa_{i,t-1}^l \quad (1 \le l \le n, 1 \le i \le q_l, 1 \le t \le L) \tag{5.5}$$

6) If O_i^l ends by t, then it ends by $t+1$:

$$eb_{i,t}^l \rightarrow eb_{i,t+1}^l \quad (1 \le l \le n, 1 \le i \le q_l, 0 \le t < L) \tag{5.6}$$

7) If O_i^l starts at or after time t, then it cannot end before time $t + p_i^l - 1$:

$$sa_{i,t}^l \rightarrow \neg eb_{i,t+p_i^l-1}^l \quad (1 \le l \le n, 1 \le i \le q_l, 0 \le t < L - p_i^l + 1) \tag{5.7}$$

8) If O_i^l starts at or after t and O_j^k follows O_i^l, then O_j^k cannot start until O_i^l is finished. That is, for $pr_{i,j}^{l,k}$ added by rules 1 and 2, we add the clause:

$$sa_{i,t}^l \wedge pr_{i,j}^{l,k} \rightarrow sa_{j,t+p_i^l}^k \quad (0 \le t \le L - p_i^l) \tag{5.8}$$

Let S be the SAT problem translated by the above Crawford encoding. If S is satisfiable, then JSSP can complete all the jobs by the makespan L. Let M be the truth assignment satisfying S. The start time t_i^l of each operation O_i^l is given by extracting the last $sa_{i,t}^l$, which is assigned as true in M. More precisely, t_i^l is given by t which satisfies the following expression:

$$\left(M \models sa_{i,t}^l\right) \wedge \left(\neg \exists u > t\ (M \models sa_{i,u}^l)\right)$$

where $M \models x$ indicates x is assigned true in M.

Let S_L be a SAT problem generated by the Crawford encoding under the assumption that the makespan is at most L. If we find a positive integer k such that S_{k-1} is unsatisfiable and S_k is satisfiable, then we conclude that the minimum makespan is k. Such a k divides SAT problems $S_i(1 \le i)$ into an unsatisfiable part $S_i(1 \le i < k)$ and a satisfiable part $S_i(k \le i)$.

5.3 Performance Evaluation

The SAT encoding approach was tested to solve six open JSSPs: ABZ8, ABZ9, YN1, YN2, YN3, and YN4. These problems were obtained from the ORLibrary.[1] Each ABZn consists of 20 jobs and 15 machines, while each YNn consists of 20 jobs

1 http://people.brunel.ac.uk/mastjjb/jeb/orlib/files/jobshop1.txt.

Table 5.2 Original LB and UB.

Problem	LB	UB
ABZ8	646	665
ABZ9	662	678
YN1	846	884
YN2	870	907
YN3	840	892
YN4	920	968

and 20 machines. Table 5.2 displays the best-known lower bounds (LB) and upper bounds (UB) as indicated in the literature [26–28].

For each problem, denoted as P, we generate $UB - LB + 2$ SAT instances $S^P_{LB-1}, S^P_{LB}, \dots$, and S^P_{UB} with the Crawford encoding and solve them with the SAT solver MiniSat [29]. Even though we know S^P_{LB-1} is unsatisfiable because of the LB and S^P_{UB} is satisfiable because of the UB, we solve them in order to verify the values. The solver used for evaluation is MiniSat version 2.

All experiments were conducted on the cluster machine A of Collaborative Facilities for Verification (CFV) SATSUKI in AIST (National Institute of Advanced Industrial Science and Technology). The cluster machine A consists of 112 nodes. Each node is a Xeon X5260 3.3 GHz machine with 8 GB memory on Linux 2.6.26-2-amd64.

5.3.1 Solving ABZ9 and YN1

The SAT encoding approach successfully obtained the optimal solutions for ABZ9 and YN1, which are 678 and 884, respectively. This confirms that the best-known UB of ABZ9 and YN1 are indeed optimal.

Tables 5.3 and 5.4 show the experimental results of $S^{ABZ9}_N (N = 661, \dots, 678)$ and $S^{YN1}_N (N = 875, \dots, 884)$, respectively. The second and third columns show the size of the SAT instance. The second shows the number of propositional variables and the third shows the number of clauses in the corresponding SAT instance. As N increases, the size of SAT instances increases linearly, while the CPU time increases exponentially. This reflects NP hardness of JSSP.

The only exception is the result of S^{ABZ9}_{678}, which is much easier than S^{ABZ9}_{677}. The main reason is that S^{ABZ9}_{678} is satisfiable while S^{ABZ9}_{677} is unsatisfiable. We finish solving S^{ABZ9}_{678} when a truth assignment satisfying S^{ABZ9}_{678} is found. Therefore, we need not examine the whole search space of S^{ABZ9}_{678} while we must examine the whole search space S^{ABZ9}_{677}.

Table 5.3 Results of $S_N^{ABZ9}(N = 661, \ldots, 678)$.

N	No. of variables	No. of clauses	CPU (seconds)	Satisfiable
661	403 180	4 402 236	6595	No
662	403 780	4 409 116	8946	No
663	404 380	4 415 996	10 363	No
664	404 980	4 422 876	12 673	No
665	405 580	4 429 756	14 279	No
666	406 180	4 436 636	16 616	No
667	406 780	4 443 516	21 905	No
668	407 380	4 450 396	26 847	No
669	407 980	4 457 276	39 692	No
670	408 580	4 464 156	42 939	No
671	409 180	4 471 036	64 439	No
672	409 780	4 477 916	104 448	No
673	410 380	4 484 796	119 798	No
674	410 980	4 491 676	132 638	No
675	411 580	4 498 556	240 493	No
676	412 180	4 505 436	268 354	No
677	412 780	4 512 316	457 021	No
678	413 380	4 519 196	10 530	Yes

Table 5.4 Results of $S_N^{YN1}(N = 875, \ldots, 884)$.

N	No. of variables	No. of clauses	CPU (seconds)	Satisfiable
875	708 780	7 799 938	184 492	No
876	709 580	7 809 118	296 314	No
877	710 380	7 818 298	373 687	No
878	711 180	7 827 478	549 011	No
879	711 980	7 836 658	620 052	No
880	712 780	7 845 838	1 058 190	No
881	713 580	7 855 018	1 166 480	No
882	714 380	7 864 198	2 007 540	No
883	715 180	7 873 378	1 912 690	No
884	715 980	7 882 558	3 149 770	Yes

Table 5.5 Improvement of LB and UB.

Problem	SAT instance	CPU (seconds)	Satisfiable	New LB	New UB
ABZ8	S_{658}^{ABZ8}	7 611 330	No	659	—
YN2	S_{892}^{YN2}	5 739 010	No	893	—
YN2	S_{905}^{YN2}	803 983	Yes	—	905
YN3	S_{879}^{YN3}	3 419 330	No	880	—
YN4	S_{947}^{YN4}	3 554 680	No	948	—

Table 5.6 New LB and UB.

Problem	LB	UB
ABZ8	659[a)]	665
ABZ9	678[b)]	
YN1	884[b)]	
YN2	893[a)]	905[c)]
YN3	880[a)]	892
YN4	948[a)]	968

a) The new LB found by SAT.
b) The optimum solution found by SAT.
c) The new UB found by SAT.

5.3.2 Improving LB and UB

The LB for ABZ8, YN2, YN3, and YN4 have been improved, along with the UB for YN2. Table 5.5 displays experimental results for the SAT instances proof of which gives the best LB or UB. A slash in the table indicates that the corresponding bound was not updated by the SAT encoding approach. Solving more difficult problems, for example, S_{895}^{YN2} and S_{950}^{YN4}, needs more than 8 GB memory which exceeds the capabilities of the machine. Table 5.6 summarizes the new LB and UB.

5.4 Chapter Summary

This chapter aimed to address six open JSSPs using the SAT encoding approach. Notably, we achieved new optimal solutions for ABZ9 and YN1, with values of 678 and 884, respectively. Additionally, we obtained new LB for all other problems and achieved a new UB for YN2. These results underscore the effectiveness of the SAT encoding approach in solving open combinatorial problems.

However, four problems (ABZ8, YN2, YN3, and YN4) still remain open. Solving these problems needs more than several months and 8 GB of memory if we follow the current approach. To address these challenging problems, parallelizing SAT solving appears to be a promising avenue.

References

1 M. L. Pinedo. *Scheduling: Theory, Algorithms, and Systems* - Sixth Edition, Springer, 2022.

2 J. Y. Lee. A genetic algorithm for a two-machine flowshop with a limited waiting time constraint and sequence-dependent setup times. *Mathematical Problems in Engineering*, 2020: 1–13, 2020.

3 Q. Jiang, X. Liao, R. Zhang, and Q. Lin. Energy-saving production scheduling in a single-machine manufacturing system by improved particle swarm optimization. *Mathematical Problems in Engineering*, 2020: 1–16, 2020.

4 H. Wei, S. Li, H. Quan, D. Liu, S. Rao, C. Li, and J. Hu. Unified multi-objective genetic algorithm for energy efficient job shop scheduling. *IEEE Access*, 9:54542–54557, 2021.

5 A. Malik, C. Walker, M. O'Sullivan, and O. Sinnen. Satisfiability modulo theory (SMT) formulation for optimal scheduling of task graphs with communication delay. *Computers and Operations Research*, 89:113–126, 2018.

6 S. A. Kravchenko and F. Werner. Parallel machine problems with equal processing times: A survey. *Journal of Scheduling*, 14(5):435–444, 2011.

7 M. O. Adamu and A. O. Adewumi. A survey of single machine scheduling to minimize weighted number of tardy jobs. *Journal of Industrial and Management Optimization*, 10(1):219–241, 2013.

8 S. Venugopalan and O. Sinnen. ILP formulations for optimal task scheduling with communication delays on parallel systems. *IEEE Transactions on Parallel and Distributed Systems*, 26(1):142–151, 2014.

9 K. Ying, C. Cheng, S. Lin, and C. Hung. Comparative analysis of mixed integer programming formulations for single-machine and parallel-machine scheduling problems. *IEEE Access*, 7:152998–153011, 2019.

10 Z. Yang, H.-S. Xiao, R. Guan, Y. Yang, and H.-L. Ji. Task scheduling for multiunit parallel test using mixed-integer linear programming. *Mathematical Problems in Engineering*, 2021:1–13, 2021.

11 M. Bofill, J. Coll, G. Martín, J. Suy, and M. Villaret. The sample analysis machine scheduling problem: Definition and comparison of exact solving approaches. *Computers & Operations Research*, 142:105730, 2022.

12 Z. Cheng, H. Zhang, Y. Tan, and Y. Lim. SMT-based scheduling for overloaded real-time systems. *IEICE Transactions on Information and Systems*, E100-D(5):1055–1066, 2017.

13 M. Bofill, J. Coll, J. Suy, and M. Villaret. SMT encodings for resource-constrained project scheduling problems. *Computers & Industrial Engineering*, 149:106777, 2020.

14 D. Lee, B. Lin, and C.-K. Cheng. SMT-based contention-free task mapping and scheduling on 2D/3D SMART NoC with mixed dimension-order routing. *ACM Transactions on Architecture and Code Optimization*, 19(1):5:1–5:21, 2022.

15 J. M. Crawford and A. B. Baker. Experimental results on the application of satisfiability algorithms to scheduling problems. In *12th AAAI National Conference on Artificial Intelligence*, pages 1092–1097, 445 Burgess Drive Menlo Park, CA, United States, October 1994. American Association for Artificial Intelligence.

16 M. Koshimura, H. Nabeshima, H. Fujita, and R. Hasegawa. Solving open job-shop scheduling problems by sat encoding. *IEICE Transactions on Information and Systems*, E93-D(8):2316–2318, 2010.

17 W. Liu, Z. Gu, J. Xu, X. Wu, and Y. Ye. Satisfiability modulo graph theory for task mapping and scheduling on multiprocessor systems. *IEEE Transactions on Parallel and Distributed Systems*, 22(8):1382–1389, 2011.

18 R. Nieuwenhuis, A. Oliveras, E. Rodríguez-Carbonell, and E. Rollon. Employee scheduling with sat-based pseudo-Boolean constraint solving. *IEEE Access*, 9:142095–142104, 2021.

19 M. Bofill, M. Garcia, J. Suy, and M. Villaret. MaxSAT-based scheduling of B2B meetings. In L. Michel, editor, *Integration of AI and OR Techniques in Constraint Programming*, pages 65–73. Springer International Publishing, 2015.

20 X. Liao, H. Zhang, M. Koshimura, R. Huang, and W. Yu. Maximum satisfiability formulation for optimal scheduling in overloaded real-time systems. In *PRICAI 2019: Trends in Artificial Intelligence - 16th Pacific Rim International Conference on Artificial Intelligence*, Cuvu, Yanuca Island, Fiji, August 26–30, 2019. *Proceedings, Part I*, volume 11670 of *Lecture Notes in Computer Science*, pages 618–631. Springer, 2019.

21 E. Demirović, N. Musliu, and F. Winter. Modeling and solving staff scheduling with partial weighted MaxSAT. *Annals of Operations Research*, 275(1):79–99, 2019.

22 X. Liao, H. Zhang, M. Koshimura, R. Huang, W. Yu, and F. Li. Modeling and solving scheduling in overloaded situations with weighted partial MaxSAT. *Mathematical Problems in Engineering*, 2021:1–17, 2021.

23 M. Bofill, J. Coll, M. Garcia, J. Giráldez-Cru, G. Pesant, J. Suy, and M. Villaret. Constraint solving approaches to the business-to-business meeting scheduling problem. *Journal of Artificial Intelligence Research*, 74:263–301, 2022.

24 J. Adams, E. Balas, and D. Zawack. The shifting bottleneck procedure for job shop scheduling. *Management Science*, 34(3):391–401, 1988.

25 T. Yamada and R. Nakano. A genetic algorithm applicable to largescale job-shop problems. In *2nd International Conference on Parallel Problem Solving from Nature*, pages 281–290, 1992.

26 A. S. Jain and S. Meeran. Deterministic job-shop scheduling: Past, present and future. *European Journal of Operational Research*, 113(2):390–434, 1999.

27 W. Brinkkötter and P. Brucker. Solving open benchmark instances for the job-shop problem by parallel head-tail adjustments. *Journal of Scheduling*, 4(1):53–64, 2001.

28 C. Zhang, P. Li, Y. Rao, and Z. Guan. A very fast TS/SA algorithm for the job shop scheduling problem. *Computers and Operations Research*, 35(1):282–294, 2008.

29 N. Eén and N. Sörensson. MiniSat: A sat solver with conflict-clause minimization. In *8th International Conference on Theory and Applications of Satisfiability Testing*, pages 502–*518*, 2005.

6
Overloaded Scheduling

After reading this chapter, you should be able to:

- Understand the overloaded scheduling model for a single machine.
- Grasp the process of encoding an overloaded scheduling problem for a single machine into a weighted partial MaxSAT (WPM) problem.
- Comprehend the method of extending the WPM encoding of single-machine scheduling to parallel-machine scheduling.

6.1 Overloaded Scheduling Model

Real-time systems, which are designed to handle tasks with completion deadlines, play an important role in a variety of modern applications, such as robotics [1], pacemakers [2], chemical plants [3], telecommunications [4], and multimedia systems [5]. Under ideal circumstances, a real-time system completes all tasks before their deadlines expire. However, in reality, the workload may exceed the system's capacity, leading to missed deadlines [6]. Such a phenomenon is referred to as overload, and the tasks that miss their deadlines are termed late or overdue tasks. A classic example is a switch in a communication network which polls its incoming links to forward packets that have arrived since its previous servicing of the link [7]. On each link, different packets may have different processing times, deadlines and importance values. When packages flood the switch, overload happens. In this situation, designing a suitable scheduling strategy to maximize the total value of forwarded packages is critical to maintain a service's stability.

To achieve the goal of minimizing the number of late tasks, Cheng et al. [8, 9] encoded the problem as a set of first-order formulas tackled by a satisfiability modulo theory (SMT) solver called Z3, showing that SMT-based scheduling is sufficiently flexible because it handles various task properties and objectives with

Applied Satisfiability: Cryptography, Scheduling, and Coalitional Games, First Edition. Xiaojuan Liao and Miyuki Koshimura.

very few changes in the adaption procedure. For example, when tasks have different importance values, and the objective turns to minimizing the total weight of late tasks, only the target constraints need to be modified. Wang et al. [10] enhanced Cheng's SMT formulation by removing redundant constraints and eliminating successive calls of the Z3 solver, thus improved the efficiency by more than two orders of magnitude. Liao et al. [11, 12] encoded both the unweighted and weighted versions of the same scheduling problem into Boolean propositional logic and showed that weighted partial MaxSAT (WPM) solvers provide a competitive alternative to SMT solvers. This chapter presents the WPM formulation for the optimal resolution of single-machine scheduling in overloaded scenarios, with the objective of minimizing the total weight value of overdue tasks. The minimization objective is equivalent to maximizing the total weight of tasks meeting their deadlines.

The definition of the overloaded scheduling problem is provided in [9, 10, 12]. For convenience, the notations used in the model are summarized in Table 6.1. The scheduling problem involves n tasks $\Gamma = \{\tau_1, \dots, \tau_n\}$ to be processed. All the tasks request a uniprocessor for execution when they arrive in the system. Each task $\tau_\ell \in \Gamma$ is represented by a 4-tuple $\tau_\ell = (r_\ell, c_\ell, d_\ell, w_\ell)$, where r_ℓ, c_ℓ, d_ℓ, and w_ℓ are all non-negative integers representing the release date, the execution time, the deadline, and the weight of τ_ℓ, respectively. Naturally, $r_\ell + c_\ell \leq d_\ell$. Weight w_ℓ reflects the importance of task τ_ℓ. The larger w_ℓ is, the more important τ_ℓ is. Given

Table 6.1 Notations and descriptions in scheduling model.

Notation	Description
Γ	Finite set of real-time tasks
n	Number of real-time tasks. $n = \lvert\Gamma\rvert$
τ_ℓ	Task in Γ, where ℓ is its index
r_ℓ	Release date of τ_ℓ
c_ℓ	Execution time of τ_ℓ
d_ℓ	Due date of τ_ℓ
w_ℓ	Weight of τ_ℓ
f_i^ℓ	ith fragment of τ_ℓ
q_ℓ	Number of fragments in τ_ℓ
c_i^ℓ	Execution time of f_i^ℓ
$\tau_k \prec \tau_\ell$	τ_ℓ succeeds τ_k
R_{dp}	Set of task pairs that have dependency relations

deadline d_ℓ, if task τ_ℓ is completed at or before d_ℓ, τ_ℓ is on time and weight w_ℓ is obtained by the system. Otherwise, τ_ℓ is late and worthless to the system.

To allow preemption, which indicates that a running task may be interrupted and resumed later, each task $\tau_\ell \in \Gamma$ is split into q_ℓ non-preemptive subtasks (fragments), and is defined as a chain of indivisible fragments $\langle f_1^\ell, \ldots, f_{q_\ell}^\ell \rangle$. Symbol c_i^ℓ stands for the required execution time of f_i^ℓ. Clearly, $\sum_{i=1}^{q_\ell} c_i^\ell = c_\ell$. For $2 \leq i \leq q_\ell$, f_i^ℓ can only start to run after f_{i-1}^ℓ is completed.

In practical systems, tasks usually have dependency relations. For example, if task τ_ℓ requires the computed result of τ_k, τ_ℓ cannot start until τ_k is finished. Such a dependency relation between tasks is written as $\tau_k \prec \tau_\ell$, where τ_k is the immediate predecessor of τ_ℓ, and τ_ℓ is the immediate successor of τ_k. Obviously, if $\tau_k \prec \tau_\ell$, the constraint $r_k + c_k \leq d_\ell - c_\ell$ should be satisfied, otherwise, τ_ℓ can never be completed no matter when its predecessor finishes. The set of task pairs that have dependency relations over Γ is denoted by $R_{dp} = \{(\tau_k, \tau_\ell) \mid \tau_k, \tau_\ell \in \Gamma, \tau_k \prec \tau_\ell\}$.

6.2 Weighted Partial MaxSAT Approach

This section provides the WPM formulation for solving the overloaded scheduling problem on a single machine. An overview of the WPM formulation is depicted in Figure 6.1.

Feature preprocessing identifies two types of constraints in the scheduling problem: constraints on scheduling features and constraints on objectives. Given a set of tasks $\Gamma = \{\tau_1, \ldots, \tau_n\}$, scheduling features uniquely characterize the problem over Γ and the objective is to seek for a schedule that completes all

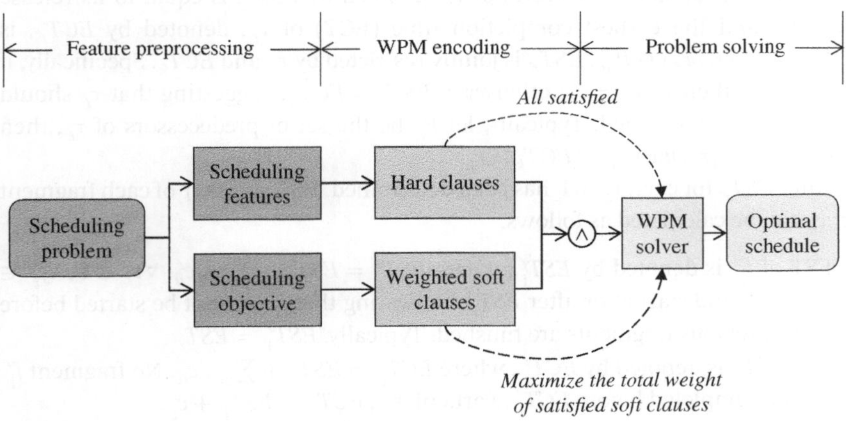

Figure 6.1 Overview of the WPM-based scheduling method (Source: Xiaojuan Liao et al., 2021/Hindawi/CC BY 4.0).

tasks on time subject to constraints on the scheduling features. In overloaded situations, making all the tasks complete by their deadlines is impossible, and the scheduling problem is then treated as an optimization problem, which aims to maximize the total weight of tasks that are completed by their deadlines. After feature preprocessing, WPM encoding can be implemented separately on the scheduling features and objective. Specifically, scheduling features, which are intrinsic and determined when tasks are released, are translated into a set of hard clauses that must be satisfied without exception. Conversely, the objective in overloaded situations is encoded as a set of weighted soft clauses that are allowed to remain unsatisfied. By combining these hard clauses with weighted soft ones, the problem transforms into a WPM problem that can be addressed by any off-the-shelf WPM solver. During the problem-solving phase, a WPM solver attempts to satisfy all the hard clauses and maximize the total weight of satisfied soft clauses. The output of the WPM solver includes the assignment of all Boolean variables, from which the optimal schedule can be derived.

Section 6.2.1 shows how to identify fragments' critical time points to characterize the scheduling problem, paving the way for the WPM encoding described in Section 6.2.2. Section 6.2.3 exhibits an example to show how the scheduling problem is addressed with the presented WPM encoding. Details on the problem-solving phase are also exhibited in Section 6.2.3.

6.2.1 Feature Preprocessing

Given a task set $\Gamma = \{\tau_1, \dots, \tau_n\}$ where each task $\tau_\ell \in \Gamma$ is characterized by $(r_\ell, c_\ell, d_\ell, w_\ell)$, $\forall \tau_\ell \in \Gamma$, τ_ℓ cannot start before its release date r_ℓ. Thus, the possible earliest start time (EST) of τ_ℓ, denoted by EST_ℓ, is equal to its release date r_ℓ, and the earliest completion time (ECT) of τ_ℓ, denoted by ECT_ℓ, is $EST_i + c_i$. $\forall(\tau_k, \tau_\ell) \in R_{dp}$, EST_ℓ is jointly restricted by r_ℓ and ECT_k. Specifically, if $ECT_k \leq r_\ell$, then $EST_\ell = r_\ell$; otherwise, $EST_\ell = ECT_k$, suggesting that τ_ℓ should wait until τ_k is finished. Typically, let P_ℓ be the set of predecessors of τ_ℓ, then $EST_\ell = max\{r_\ell, max_{\tau_k \in P_\ell}\{ECT_k\}\}$.

After EST_ℓ for each τ_ℓ in Γ has been determined, EST and ECT of each fragment in τ_ℓ can be calculated as follows:

- EST of f_i^ℓ is denoted by EST_i^ℓ, where $EST_i^\ell = EST_\ell + \sum_{u=1}^{i-1} c_u^\ell$. $\forall \tau_\ell \in \Gamma$, $\forall f_i^\ell \in \tau_\ell$, f_i^ℓ should start at or after EST_i^ℓ, indicating that f_i^ℓ cannot be started before all its previous fragments are finished. Typically, $EST_1^\ell = EST_\ell$.
- ECT of f_i^ℓ is denoted by ECT_i^ℓ, where $ECT_i^\ell = EST_\ell + \sum_{u=1}^{i} c_u^\ell$. No fragment f_i^ℓ can be completed before ECT_i^ℓ. Particularly, $ECT_i^\ell = EST_i^\ell + c_i^\ell$.

In addition to EST and ECT, $\forall \tau_\ell \in \Gamma$, each fragment $f_i^\ell \in \tau_\ell$ is characterized by the following two types of time:

- Latest start time (LST) of f_i^ℓ is denoted by LST_i^ℓ, where $LST_i^\ell = d_\ell - \sum_{u=i}^{q_\ell} c_u^\ell$. If f_i^ℓ fails to start before LST_i^ℓ, τ_ℓ cannot end by its deadline d_ℓ, and thus this task becomes worthless to the system.
- Latest completion time (LCT) of f_i^ℓ is denoted by LCT_i^ℓ, where $LCT_i^\ell = d_\ell - \sum_{u=i+1}^{q_\ell} c_u^\ell$. Particularly, $LCT_i^\ell = LST_i^\ell + c_i^\ell$.

Note that in the above four critical time points, ECT is unrelated to the WPM encoding. Nevertheless, it is indispensable in the previous encoding [11], which will be introduced in Section 6.3.

6.2.2 WPM Formulation

This section introduces how to encode the scheduling problem as a WPM problem. A WPM instance consists of a number of clauses that need to be managed by the WPM solver. To formulate all the necessary constraints that characterize the scheduling model, we introduce the following three Boolean variables:

- $sa_{i,t}^\ell$, which is true if f_i^ℓ starts at time t or later;
- $pr_{i,j}^{\ell,k}$, which is true if f_i^ℓ precedes f_j^k;
- eb_ℓ, which is true if τ_ℓ ends by its deadline d_ℓ.

Based on the generated Boolean variables, the WPM encoding can be implemented. In what follows, several rules are presented to encode the features of fragments into a set of hard clauses. The main encoding is derived and enhanced from previous work [11, 13, 14]. Logical implication $a \rightarrow b$ is equivalent to $\neg a \vee b$ in classical logic.

(C1) $\forall \tau_\ell \in \Gamma$, the first fragment of τ_ℓ, i.e., f_1^ℓ, starts at or after EST_1^ℓ:

$$sa_{1,EST_1^\ell}^\ell \tag{6.1}$$

(C2) $\forall \tau_\ell \in \Gamma, \forall f_i^\ell, f_{i+1}^\ell \in \tau_\ell, f_i^\ell$ precedes f_{i+1}^ℓ:

$$pr_{i,i+1}^{\ell,\ell} \tag{6.2}$$

(C3) $\forall \tau_k, \tau_\ell \in \Gamma$, $\forall f_i^k \in \tau_k$, $\forall f_j^\ell \in \tau_\ell$, if $k \neq l$, $\tau_k \nprec \tau_l$, $\tau_l \nprec \tau_k$, $EST_i^k < LCT_j^\ell$ and $EST_j^\ell < LCT_i^k$, and then f_i^k and f_j^ℓ may require the processor at the same time. In this condition, f_i^k precedes f_j^ℓ or f_j^ℓ precedes f_i^k:

$$pr_{i,j}^{k,\ell} \vee pr_{j,i}^{\ell,k} \tag{6.3}$$

(C4) $\forall (\tau_k, \tau_\ell) \in R_{dp}$, if τ_k fails to be completed by its deadline, then τ_ℓ cannot even start at LST_1^ℓ:

$$\neg eb_k \rightarrow sa_{1,LST_1^\ell+1}^\ell \tag{6.4}$$

Furthermore, if $d_k > EST_1^\ell$, the first fragment of τ_ℓ cannot start until the last fragment of τ_k finishes. That is, if $d_k > EST_1^\ell$, then the constraint that $f_{q_k}^k$ precedes f_1^ℓ is enforced:

$$pr_{q_k,1}^{k,\ell} \tag{6.5}$$

(C5) $\forall \tau_\ell \in \Gamma, \forall f_i^\ell \in \tau_\ell$, if f_i^ℓ starts at or after time t, then it starts at or after time $t-1$:

$$sa_{i,t}^\ell \rightarrow sa_{i,t-1}^\ell \quad \left(EST_i^\ell + 1 \leq t \leq LST_i^\ell + 1\right) \tag{6.6}$$

(C6) $\forall \tau_\ell \in \Gamma$, if τ_ℓ ends before its deadline d_ℓ, then last fragment $f_{q_\ell}^\ell$ must start at or before time $LST_{q_\ell}^\ell$. In other words, $f_{q_\ell}^\ell$ cannot start at or after time $LST_{q_\ell}^\ell + 1$:

$$eb_\ell \rightarrow \neg sa_{q_\ell, LST_{q_\ell}^\ell + 1}^\ell \tag{6.7}$$

(C7) $\forall \tau_k, \tau_\ell \in \Gamma, \forall f_i^k \in \tau_k, \forall f_j^\ell \in \tau_\ell$, if f_i^k starts at or after time t and f_j^ℓ follows f_i^k, then f_j^ℓ cannot start until f_i^k is finished. That is, for each $pr_{i,j}^{k,\ell}$ asserted by (C2)–(C4), one clause is generated:

$$sa_{i,t}^k \wedge pr_{i,j}^{k,\ell} \rightarrow sa_{j,t'}^\ell \tag{6.8}$$

where t varies in $[EST_i^k, LST_i^k + 1]$ and

$$t' = \begin{cases} LST_j^\ell + 1 & \text{if} \quad t + c_i^k > LST_j^\ell \\ EST_j^\ell & \text{if} \quad t + c_i^k < EST_j^\ell \\ t + c_i^k & \text{Otherwise} \end{cases} \tag{6.9}$$

This formula reveals the following facts. First, if f_i^k ends after LST_j^ℓ (i.e., $t + c_i^k > LST_j^\ell$), then f_j^ℓ cannot start at or before LST_j^ℓ. Second, if f_i^k finishes before EST_j^ℓ (i.e., $t + c_i^k < EST_j^\ell$), then f_j^ℓ starts at or after EST_j^ℓ. Otherwise, f_j^ℓ must start at or after time f_i^k finishes, i.e., $t + c_i^k$.

Up to this point, we have encoded the scheduling features as Boolean formulas that can be converted to a set of clauses. Since the scheduling features are intrinsic properties inherent in the tasks and their fragments, such clauses are specified as hard, indicating that all of them must absolutely be satisfied. For convenience, we refer to the set of hard clauses introduced in (C1)–(C7) as $\mathscr{C}$.

A task is said to be on-time if and only if it is completed before its deadline. Thus the scheduling objective can be directly encoded by the following rule:

(O) Maximizing the sum of the weights of on-time tasks:

$$\left(eb_\ell, w_\ell\right) \quad (1 \leq \ell \leq n) \tag{6.10}$$

Equation (6.10) indicates that if clause eb_ℓ is satisfied (i.e., evaluating to True), then weight w_ℓ is gained; otherwise, the gain is zero. To simplify the discussion, we introduce $\mathcal{O}$ to denote the set of clauses in Eq. (6.10). In an overloaded system, not all tasks can be completed before their deadlines. To handle such situations, we declare the clauses in $\mathcal{O}$ to be soft, indicating that completing all the tasks by their deadlines is a soft constraint. Conjuncted with $\mathcal{C}$, the problem is then $\{\mathcal{C}, \mathcal{O}\}$. This leads to a WPM problem, which tries to find an assignment of variables to satisfy all the hard clauses in $\mathcal{C}$ and to maximize the sum of the weights of the satisfied soft clauses in $\mathcal{O}$, i.e., to maximize $\sum_{\ell=1}^{n} w_\ell \cdot eb_\ell$.

6.2.3 A Pedagogical Example

Let us consider a simple scheduling problem to describe how WPM formulation works. As shown in Figure 6.2a, there is a set of real-time tasks $\Gamma = \{\tau_1, \tau_2, \tau_3\}$, where both τ_1 and τ_3 rely on the computed result of τ_2, represented as $\tau_2 \prec \tau_1$ and $\tau_2 \prec \tau_3$, i.e., $R_{dp} = \{(\tau_2, \tau_1), (\tau_2, \tau_3)\}$. The release dates, the execution times, the due dates, and the weights of these tasks are respectively defined as $\tau_1 = (0, 2, 3, 1)$, $\tau_2 = (0, 1, 1, 2)$, $\tau_3 = (0, 1, 2, 3)$. Suppose that τ_1 has two fragments, i.e., $\langle f_1^1, f_2^1 \rangle$. τ_2 and τ_3 each have one, denoted by f_1^2 and f_1^3, respectively. The execution time of each fragment is 1.

The refined problem exhibition and the critical time points after feature preprocessing are summarized in Figure 6.2b and Table 6.2, respectively. The MaxSAT formulation applied to the overloaded scheduling problem is shown

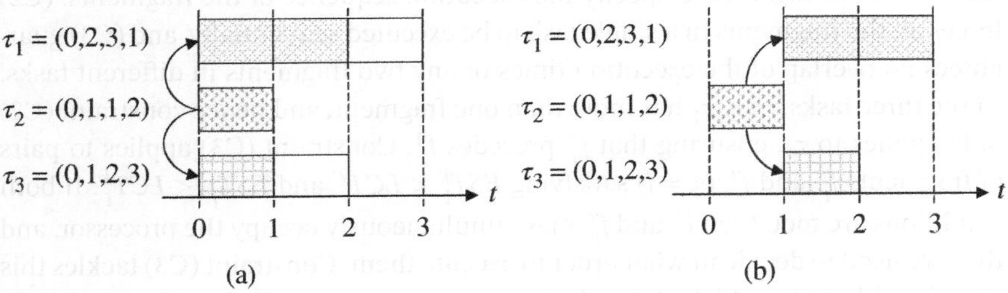

Figure 6.2 A simple scheduling example. (a) Original problem and (b) problem after feature preprocessing (Source: Xiaojuan Liao et al., 2021/Hindawi/CC BY 4.0).

Table 6.2 Critical time points of each fragment in three tasks.

Task	Fragment	EST	LST	LCT
τ_1	f_1^1	1	1	2
	f_2^1	2	2	3
τ_2	f_1^2	0	0	1
τ_3	f_1^3	1	1	2

$$
\begin{array}{ll}
(\mathrm{C1}) & sa_{1,1}^{1} \\
\wedge & sa_{1,0}^{2} \\
\wedge & sa_{1,1}^{3} \\
\wedge & \\
(\mathrm{C2}) & pr_{1,2}^{1,1} \\
\wedge & \\
(\mathrm{C3}) & pr_{1,1}^{1,3} \vee pr_{1,1}^{3,1} \\
\wedge & \\
(\mathrm{C4}) & eb_2 \vee sa_{1,2}^{1} \\
\wedge & eb_2 \vee sa_{1,2}^{3} \\
(\mathrm{C5}) & \neg sa_{1,2}^{1} \vee sa_{1,1}^{1} \\
\wedge & \neg sa_{2,3}^{1} \vee sa_{2,2}^{1} \\
\wedge & \neg sa_{1,1}^{2} \vee sa_{1,0}^{2} \\
\wedge & \neg sa_{1,2}^{3} \vee sa_{1,1}^{3} \\
\wedge &
\end{array}
\qquad
\begin{array}{ll}
(\mathrm{C6}) & \neg sa_{2,3}^{1} \vee \neg eb_1 \\
\wedge & \neg sa_{1,1}^{2} \vee \neg eb_2 \\
\wedge & \neg sa_{1,2}^{3} \vee \neg eb_3 \\
\wedge & \\
(\mathrm{C7}) & \neg sa_{1,1}^{1} \vee \neg pr_{1,2}^{1,1} \vee sa_{2,2}^{1} \\
\wedge & \neg sa_{1,2}^{1} \vee \neg pr_{1,2}^{1,1} \vee sa_{2,3}^{1} \\
\wedge & \neg sa_{1,1}^{1} \vee \neg pr_{1,1}^{1,3} \vee sa_{1,2}^{3} \\
\wedge & \neg sa_{1,2}^{1} \vee \neg pr_{1,1}^{1,3} \vee sa_{1,2}^{3} \\
\wedge & \neg sa_{1,1}^{3} \vee \neg pr_{1,1}^{3,1} \vee sa_{1,2}^{1} \\
\wedge & \neg sa_{1,2}^{3} \vee \neg pr_{1,1}^{3,1} \vee sa_{1,2}^{1} \\
\wedge & \\
(\mathrm{O}) & (eb_1, 1) \\
\wedge & (eb_2, 2) \\
\wedge & (eb_3, 3)
\end{array}
$$

Figure 6.3 WPM formulation for exemplified problem (Source: Xiaojuan Liao et al., 2021/Hindawi/CC BY 4.0).

in Figure 6.3. For the complete file in WPM input format, readers may refer to Appendix 14.

Constraint (C1) states that each task starts at or after its EST. Constraints (C2) and (C3) work together to specify the execution sequence of the fragments. (C2) forces all the fragments in a single task to be executed sequentially, and (C3) guarantees no overlap of the execution times of any two fragments in different tasks. In the three tasks, only τ_1 has more than one fragment, and hence constraint (C2) only applies to τ_1, ensuring that f_1^1 precedes f_2^1. Constraint (C3) applies to pairs of fragments f_i^k and f_j^ℓ $(k \neq l)$ satisfying $EST_i^k < LCT_j^\ell$ and $EST_j^\ell < LCT_i^k$. If both conditions are met, then f_i^k and f_j^ℓ may simultaneously occupy the processor, and thus we need to decide in what order to execute them. Constraint (C3) tackles this ordering dilemma, which states that either one can precede the other. Consider f_1^1 and f_1^3. As seen in Table 6.2, $EST_1^1 < LCT_1^3$ and $EST_1^3 < LCT_1^1$, hence we must explicitly specify that f_1^1 precedes f_1^3 or f_1^3 precedes f_1^1; otherwise, the execution time of these two fragments may overlap. Constraint (C4) applies to a situation where tasks have dependency relations. In the example, τ_2 is the predecessor of both τ_1 and τ_3, thus we should give up processing τ_1 and τ_3 if τ_2 misses its due date. In addition, given $d_2 = EST_1^1$ and $d_2 = EST_1^3$, there is no need to explicitly specify the execution sequence of the predecessor and successors since τ_2 must have finished before τ_1 and τ_3 start. Constraints (C5)–(C7) are partially extracted from a collection of coherence conditions [13] on the introduced variables for all

$$
\begin{array}{llll}
 & sa_{1,1}^1 = 1 & sa_{1,0}^2 = 1 & \\
eb_1 = 0 & sa_{1,2}^1 = 1 & sa_{1,1}^2 = 0 & pr_{1,2}^{1,1} = 1 \\
eb_2 = 1 & sa_{2,2}^1 = 1 & sa_{1,1}^3 = 1 & pr_{1,1}^{1,3} = 0 \\
eb_3 = 1 & sa_{2,3}^1 = 1 & sa_{1,2}^3 = 0 & pr_{1,1}^{3,1} = 1
\end{array}
$$

Figure 6.4 Assignment of all Boolean variables in the exemplified problem (Source: Xiaojuan Liao et al., 2021/Hindawi/CC BY 4.0).

the fragments of all the tasks. Finally, constraint (O) gives the problem's objective, i.e., completing the last fragment of each task by its due date.

All the clauses are conjuncted with $\wedge$ to form a WPM problem in Conjunctive Normal Form (CNF), where clauses (C1)–(C7) are declared hard and those in (O) are soft. Then the CNF formula is input to a WPM solver. The solver's output includes the maximum sum of the weights of the satisfied soft clauses as well as the corresponding assignment of all the Boolean variables.

In the exemplified problem, the assigned truth values of all the Boolean variables are listed in Figure 6.4, from which the exact start time of each fragment can be derived. Consider f_1^2. $sa_{1,0}^2 = 1$ and $sa_{1,1}^2 = 0$ indicate that f_1^2 starts at or after time 0, but it does not start at or after time 1. Then we can readily determine that f_1^2 starts at time 0. A similar reference can be made on f_1^3, which starts at time 1. Both τ_2 and τ_3 can be completed by their due dates. This is ensured by $eb_2 = 1$ and $eb_3 = 1$. Now, consider f_1^1. Figure 6.4 shows that $sa_{1,1}^1 = 1$ and $sa_{1,2}^1 = 1$. Since $LST_1^1 = 1$, that $sa_{1,2}^1 = 1$ means that f_1^1 cannot start at or before LST_1^1. Similarly, as indicated by the fact that $sa_{2,3}^1 = 1$, f_2^1 cannot start at or before LST_2^1 either. Thus, we infer that τ_1 cannot finish on time. This is confirmed by $eb_1 = 0$. As a result, the maximized total weight of the completed on-time tasks is $2 + 3 = 5$, achieved by completing τ_2 and τ_3 by their due dates.

6.3 Theoretical Discussion

Recall that in the scheduling model described in Section 6.1, each task τ_ℓ is represented as a 4-tuple $\tau_\ell = (r_\ell, c_\ell, d_\ell, w_\ell)$. Consider a special case where the weights of all the tasks are equal to a constant w, i.e., $\forall \tau_\ell \in \Gamma$, $w_\ell = w$. According to Section 6.2.2, to encode this problem, a set of weighted soft clauses $\mathcal{O} = \{(eb_\ell, w)\}$ are introduced, where $1 \le \ell \le n$. Conjuncted with hard clauses $\mathcal{C}$, problem $\{\mathcal{C}, \mathcal{O}\}$ becomes a special WPM problem with equal weights that tries to find an assignment of variables to satisfy all hard clauses in $\mathcal{C}$ and to maximize $w \cdot \sum_{\ell=1}^{n} eb_\ell$. Obviously, the essence of maximizing $w \cdot \sum_{\ell=1}^{n} eb_\ell$ is equivalent to maximizing $\sum_{\ell=1}^{n} eb_\ell$. This leads to a Partial MaxSAT (PM) problem encoded by Liao et al. [11], where each task τ_ℓ is modeled as a 3-tuple $\tau_\ell = (r_\ell, c_\ell, d_\ell)$, and the scheduling goal is to maximize the total number of on-time tasks. This section makes theoretical comparisons between PM- [11] and the WPM-based optimization frameworks when all the tasks have equal weights.

6.3.1 Similarities of PM and WPM Formulations

Essentially, the purpose of solving the scheduling problem is to determine the start time of each fragment to construct an optimal scheduling solution. To describe such crucial information, $\forall \tau_\ell \in \Gamma$, $\forall f_i^\ell \in \tau_\ell$, a set of Boolean variables $sa_{i,t}^\ell$ ($EST_i^\ell \leq t \leq LST_i^\ell + 1$) is introduced to indicate whether f_i^ℓ starts at or after time t. As long as the truth values of sa variables are all assigned, one can readily determine the exact start time of each fragment. To be specific, if a certain time $t' \in [EST_i^\ell, LST_i^\ell]$ is found such that $sa_{i,t}^\ell = 1$ for $t \leq t'$ and $sa_{i,t}^\ell = 0$ for $t > t'$, then the start time of f_i^ℓ is t' can be determined. If $sa_{i,t}^\ell$ is constantly true when t varies in $[EST_i^\ell, LST_i^\ell + 1]$, we can conclude that τ_ℓ is late. Note that it is impossible to designate $sa_{i,t}^\ell$ as true while making $sa_{i,t-1}^\ell$ false, since if f_i^ℓ starts at or after time t, f_i^ℓ must start at or after $t-1$. This constraint is interpreted by rule (C5) in the WPM encoding. In addition, $\forall \tau_\ell \in \Gamma$, $\forall f_i^\ell \in \tau_\ell$, the start time of f_i^ℓ should be at or after EST_i^ℓ. Compared with PM, WPM encodes this constraint in a more concise way by (C1). The discussion is postponed later in Section 6.3.2.

Aside from the inherent properties of a fragment's start time specified in (C1) and (C5), whether a fragment can start at a certain time is constrained by two other factors. The first is the execution order of the fragments, which should be explicitly pointed out in the following two situations: if two fragments belong to the same task, then they should be executed sequentially; if two fragments from different tasks may simultaneously occupy the processor, then either one can precede the other. The execution order of two fragments, f_i^k and f_j^ℓ, is characterized by Boolean variable $pr_{i,j}^{k,\ell}$, and the constraints specified in the above two cases are interpreted by rules (C2) and (C3), respectively. If the execution order of the two fragments is determined, the relation of their start times should be specified. To be specific, if f_i^k precedes f_j^ℓ, then f_j^ℓ must start after f_i^k finishes. This constraint is guaranteed by rule (C7). Up to this point, all the Boolean variables and clauses introduced by PM and WPM are identical. That is, rules (C2), (C3), (C5), and (C7) are all consistent with those introduced in PM.

6.3.2 WPM Improvement

As mentioned in Section 6.3.1, $\forall \tau_\ell \in \Gamma$, $\forall f_i^\ell \in \tau_\ell$, the start time of f_i^ℓ cannot be earlier than EST_i^ℓ. PM interprets this constraint by introducing one hard clause $sa_{i,EST_i^\ell}^\ell$ for $1 \leq i \leq q_\ell$, thus a total of q_ℓ clauses are generated for τ_ℓ. In comparison, as described by (C1) in Section 6.2.2, WPM generates only one clause $sa_{1,EST_1^\ell}^\ell$

for each task $\tau_\ell \in \Gamma$ no matter how many fragments τ_ℓ contains. By combining (C1) with constraints (C2) and (C7), we can easily obtain $sa^\ell_{i,EST^\ell_i} = 1$ for $1 \leq i \leq q_\ell$, which is the same as that declared by PM. Therefore, the updated rule (C1) reduces the number of generated clauses while maintaining the correctness of the encoding.

For tasks with dependency relations, their execution sequence may be explicitly or implicitly specified. In particular, $\forall(\tau_k, \tau_\ell) \in R_{dp}$, PM indistinguishably generates $pr^{k,\ell}_{q_k,1}$ for the dependency relation $\tau_k \prec \tau_\ell$. In fact, if the due date of τ_k is no later than the EST of τ_ℓ, i.e., $d_k \leq EST^\ell_1$, τ_ℓ naturally starts after τ_k is completed, thus the constraint that explicitly imposed on the execution sequence is unnecessary and can be omitted safely. WPM formulation refined the encoding in (C4) by introducing $pr^{k,\ell}_{q_k,1}$ to interpret the constraint that τ_k precedes τ_ℓ only when $d_k > EST^\ell_1$, thus redundant variables and clauses are eliminated.

Another factor that affects a fragment's start time is the completion time of a task. In general, if a task is expected to be completed by a particular time, then each fragment of the task should not start later than its LST. The relation of completion time and LST is encoded by PM and WPM in different ways. In PM, for each fragment f^ℓ_i $(1 \leq \ell \leq n, 1 \leq i \leq q_\ell)$, a set of Boolean variables $eb^\ell_{i,t}$ $(ECT^\ell_i - 1 \leq t \leq LCT^\ell_i)$ is introduced to indicate whether f^ℓ_i finishes at or before time t. Then a set of hard clauses are generated to restrict the values of the *eb* variables:

(H1) If f^ℓ_i ends by t, then it ends by time $t + 1$:

$$eb^\ell_{i,t} \rightarrow eb^\ell_{i,t+1} \quad (ECT^\ell_i - 1 \leq t \leq LCT^\ell_i - 1) \tag{6.11}$$

(H2) if τ_ℓ is completed by its due date, then $\forall f^\ell_i \in \tau_\ell, f^\ell_i$ should finish by LCT^ℓ_i:

$$eb^\ell_{q_\ell,d_\ell} \rightarrow eb^\ell_{i,LCT^\ell_i} \quad (1 \leq i \leq q_\ell - 1) \tag{6.12}$$

(H3) if f^ℓ_i starts at or after time t, then it cannot end before time $t + c^\ell_i - 1$:

$$sa^\ell_{i,t} \rightarrow \neg eb^\ell_{i,t+c^\ell_i-1} \quad (EST^\ell_i \leq t \leq LST^\ell_i + 1) \tag{6.13}$$

In fact, not all the variables and clauses in (H1)–(H3) are requisite. Given a task, instead of generating variables to characterize each fragment's completion time, it suffices to create only one variable to declare the task's due date and introduce one clause to clarify how the task's due date both restricts and is restricted by its start time. In particular, $\forall\tau_\ell \in \Gamma$, WPM introduces Boolean variable eb_ℓ to describe whether τ_ℓ can finish by its due date. Then a constraint is set to correlate τ_ℓ's due date and its start time as follows: if τ_ℓ ends by d_ℓ, then $f^\ell_{q_\ell}$ must start at or before $LST^\ell_{q_\ell}$. In other words, if $f^\ell_{q_\ell}$ fails to start at or before $LST^\ell_{q_\ell}$, then it cannot end by d_ℓ. This constraint is encoded by just one clause described in rule (C6), as declared in Section 6.2.2:

$$eb_\ell \rightarrow \neg sa^\ell_{q_\ell,LST^\ell_{q_\ell}+1} \tag{6.14}$$

Theorem 6.1 Given task $\tau_\ell \in \Gamma$, if $eb^\ell_{q_\ell,d_\ell}$ in PM (resp. eb_ℓ in WPM) is evaluated as true, then sa^ℓ_{i,LST^ℓ_i+1} for $1 \le i \le q_\ell$ in both PM and WPM are false.

Proof: In PM, according to (H2), $eb^\ell_{q_\ell,d_\ell} = 1$ makes $eb^\ell_{i,LCT^\ell_i} = 1$ for $1 \le i \le q_\ell - 1$. This indicates that $\forall f^\ell_i \in \tau_\ell, f^\ell_i$ should finish by LCT^ℓ_i. Combined with (H3), it can be inferred that $sa^\ell_{i,LST^\ell_{i+1}} = 0$ for $1 \le i \le q_\ell$.

In WPM, that $eb_\ell = 1$ leads to $sa^\ell_{q_\ell,LST^\ell_{q_\ell}+1} = 0$ by (C6). In addition, based on (C2) and (C7), it is clear that $sa^\ell_{q_\ell-1,LST^\ell_{q_\ell-1}+1} \rightarrow sa^\ell_{q_\ell,LST^\ell_{q_\ell}+1}$ holds. Under the condition where $sa^\ell_{q_\ell,LST^\ell_{q_\ell}+1} = 0$, it can be immediately inferred that $sa^\ell_{q_\ell-1,LST^\ell_{q_\ell-1}+1} = 0$. By repeatedly substituting (C2) into (C7), we finally obtain $sa^\ell_{i,LST^\ell_{i+1}} = 0$ for $1 \le i \le q_\ell$.

Theorem 6.2 Given task $\tau_\ell \in \Gamma$, if $\exists f^\ell_i \in \tau_\ell$ that makes sa^ℓ_{i,LST^ℓ_i+1} evaluate as true, then $eb^\ell_{q_\ell,d_\ell}$ in PM (resp. eb_ℓ in WPM) is false.

Proof: In PM, by (H3), $sa^\ell_{i,LST^\ell_i+1} = 1$ makes $eb^\ell_{i,LCT^\ell_i} = 0$. Combined with (H2), it is obvious that $eb^\ell_{q_\ell,d_\ell} = 0$.

In WPM, we discuss the following two cases.

1) If critical f^ℓ_i, which makes $sa^\ell_{i,LST^\ell_i+1} = 1$, is the last fragment in τ_ℓ, i.e., $i = q_\ell$, it is obvious that $eb_\ell = 0$ by (C6).
2) If $1 \le i \le q_\ell - 1$, according to (C2) and (C7), we have $sa^\ell_{i,LST^\ell_i+1} \rightarrow sa^\ell_{i+1,LST^\ell_{i+1}+1}$. Since $sa^\ell_{i,LST^\ell_i+1} = 1$, it is clear that $sa^\ell_{i+1,LST^\ell_{i+1}+1} = 1$. The step of substituting (C2) into (C7) can be repeated until $i + 1 = q_\ell$. Finally, we have $sa^\ell_{q_\ell,LST^\ell_{q_\ell}+1} = 1$. Then by (C6), $eb_\ell = 0$ is obtained.

Theorem 6.1 shows that if $eb^\ell_{q_\ell,d_\ell} = 1$ in PM (resp. $eb_\ell = 1$ in WPM), both PM and WPM guarantee that $sa^\ell_{i,LST^\ell_{i+1}} = 0$ for $1 \le i \le q_\ell$. Theorem 6.2 shows that when there exists a fragment $f^\ell_i \in \tau_\ell$ that leads to $sa^\ell_{i,LST^\ell_i+1} = 1$, both PM and WPM come to the same conclusion that τ_ℓ cannot finish by τ_ℓ, pointed out by $eb^\ell_{q_\ell,d_\ell} = 0$ and $eb_\ell = 0$ in PM and WPM, respectively. Therefore, the *eb* variables introduced in PM and WPM have the same impact on the *sa* variables and vice versa. Note that $\forall f^\ell_{q_\ell} \in \tau_\ell$, if $sa^\ell_{q_\ell,LST^\ell_{q_\ell}+1} = 0$, a WPM solver always prefers $eb^\ell_{q_\ell,d_\ell} = 1$ in PM (resp. $eb_\ell = 1$ in WPM) to satisfy soft clause $eb^\ell_{q_\ell,d_\ell}$ (resp. (eb_ℓ, w)). In other words, a WPM solver never prefers $eb^\ell_{q_\ell,d_\ell} = 0$ in PM (resp. $eb_\ell = 0$ in WPM) as long as all the hard clauses input to the solver are satisfied.

Table 6.3 Encodings related to *eb* variables in PM and WPM formulations.

		PM [11]	WPM
Variables		$eb^1_{1,0}, eb^1_{1,1}, eb^1_{1,2}, eb^1_{2,1}, eb^1_{2,2}, eb^1_{2,3}$	eb_1
Clauses	(H1)	$\neg eb^1_{1,0} \vee eb^1_{1,1}, \neg eb^1_{1,1} \vee eb^1_{1,2}, \neg eb^1_{2,1} \vee eb^1_{2,2}, \neg eb^1_{2,2} \vee eb^1_{2,3}$	$\neg eb_1 \vee \neg sa^1_{2,3}$
	(H2)	$\neg eb^1_{2,3} \vee eb^1_{1,2}$	
	(H3)	$\neg sa^1_{1,0} \vee \neg eb^1_{1,0}, \neg sa^1_{1,1} \vee \neg eb^1_{1,1}, \neg sa^1_{1,2} \vee \neg eb^1_{1,2}, \neg sa^1_{2,1} \vee \neg eb^1_{2,1}, \neg sa^1_{2,2} \vee \neg eb^1_{2,2}, \neg sa^1_{2,3} \vee \neg eb^1_{2,3}$	

Example 6.1 This example shows the differences between the PM and WPM formulations when dealing with the completion time. Assume that $\tau_1 = (0, 2, 3, 1)$ is a task in Γ, which contains two fragments $\langle f^1_1, f^1_2 \rangle$. Each fragment has execution time 1. Table 6.3 shows the Boolean variables and the hard clauses introduced by PM and WPM to characterize the fragments' completion times.

First, we can determine that $ECT^1_1 = 1$, $LCT^1_1 = 2$, $ECT^1_2 = 2$, and $LCT^1_2 = 3$. In order to encode τ_1, PM introduces a series of Boolean variables, $\{eb^1_{1,0}, eb^1_{1,1}, eb^1_{1,2}\}$ and $\{eb^1_{2,1}, eb^1_{2,2}, eb^1_{2,3}\}$ for f^1_1 and f^1_2, respectively, representing whether each fragment is completed by time point t, where $ECT^1_i - 1 \leq t \leq LCT^1_i$ for $1 \leq i \leq 2$. The relation of $eb^1_{i,t}$ and $eb^1_{i,t+1}$ is specified by rule (H1). To make τ_1 finish by its due date, f^1_1 should be completed no later than time 2. This is guaranteed by rule (H2). Finally, the relation between a fragment's start and completion times is indicated by rule (H3). Therefore, to constrain the completion time of each fragment, six Boolean variables and 11 hard clauses are generated by PM. By contrast, WPM introduces only one variable eb_1 to indicate whether τ_1 is completed by its due date. Then a hard clause is generated to constrain τ_1's due date and its last fragment's start time. In this way, both the Boolean variables and clauses generated by the WPM formulation are reduced.

6.4 Performance Evaluation

This section scrutinizes the performance comparisons of the WPM formulation and two satisfiability-based formulations on a set of randomly generated problems. The experimental design is described in Section 6.4.1, followed by the evaluation of WPM, SMT [10], and PM formulations [11] in Sections 6.4.2 and 6.4.3. All tests were conducted on a 3.4-GHz Intel E3-1230 processor with 8-GB RAM. The selected solver to evaluate WPM and PM is QMaxSAT [15], which is a SAT-based solver using the CNF encodings of cardinality constraints. The solver for SMT is Z3 [16], which is a high-performance theorem prover chosen by the previous SMT formulations [9, 10].

6.4.1 Experimental Design

The procedure for generating test instances is similar with [9–12], with a variety of parameter values. Tasks' release times are randomly generated following a discrete uniform distribution with an arriving rate λ, which represents the number of tasks that arrive during each 100 time units. Clearly, a larger λ indicates that more tasks arrive in the system during a specific period of time, thus causing heavier overload. For each task τ_ℓ, the execution time c_ℓ and the number of fragments in τ_ℓ (denoted by q_ℓ) are also randomly generated according to a discrete uniform distribution. The value of deadline d_ℓ is calculated by the formula $d_\ell = r_\ell + sf_\ell \cdot c_\ell$, where sf_ℓ is a slack factor that reflects the tightness of the due date. Weight w_ℓ of task τ_ℓ is randomly chosen from 1 to n in Section 6.4.2 and is taken as constant 1 in Section 6.4.3. The number of rule pairs with dependency relations is set 10% to the total number of tasks.

The following factors may affect the performance of the satisfiability-based formulations, including the number of tasks n, the task density (determined by λ), the execution time c_ℓ, the number of fragments q_ℓ, and the slack factor sf_ℓ. Table 6.4 provides a summary of the experimental design, where $DU(i,j)$ means the discrete uniform distribution over integer interval $[i,j]$. For each individual scenario, 100 problem instances are generated. For each instance and solver, a time limit of 300 seconds is set. If the solver fails to output the optimal result of

Table 6.4 Experimental design.

S	n	λ	c_ℓ	q_ℓ	sf_ℓ
S1	50, …, 300	10	$DU(1,30)$	$DU(1,3)$	$DU(1,4)$
S2	100	5, …, 25	$DU(1,30)$	$DU(1,3)$	$DU(1,4)$
S3	100	10	$DU(1,\alpha)$ $\alpha = 10,\dots,100$	$DU(1,3)$	$DU(1,4)$
S4	100	10	$DU(1,30)$	$DU(1,\beta)$ $\beta = 2,\dots,12$	$DU(1,4)$
S5	100	10	$DU(1,30)$	$DU(1,3)$	$DU(1,\gamma)$ $\gamma = 1,\dots,7$

an instance within the time limit, we terminate the procedure and move to the next problem instance.

6.4.2 Comparison on Weighted Cases

This section evaluates the behavior of WPM and SMT formulations [10] for maximizing the total weight value of on-time tasks. The main metrics for comparison include (1) the proportion of instances solved within the given time limit and (2) the average computation time spent on solved instances. The performance of these two formulations in different types of scenarios are shown in Figure 6.5. Each data point is the average computation time of the solved problem instances. A number with an arrow in the figures denotes the proportion of instances solved within the time limit and is omitted if the solver addresses all the 100 instances. When the proportion of the solved instances drops to zero, the corresponding curve is omitted.

As shown in Figure 6.5, both formulations perform worse as the values of all parameters increase. However, in all the varied scenarios, WPM substantially outperforms SMT in terms of both the average computation time and the proportion of successfully solved instances. This demonstrates the superiority of WPM in solving the overloaded scheduling problems.

6.4.3 Comparison on Unweighted Cases

A partial MaxSAT (PM) formulation for optimal scheduling on a single machine was presented in [11]. The scheduling objective tackled by PM differs from the WPM formulation, in that PM aims to maximize the total number of on-time tasks, while WPM tries to maximize the total weight of the on-time tasks. When all the tasks have equal weights, these two objectives become identical, and thus, formulations on these two objectives are directly comparable. This section compares the SMT [10], PM [11], and WPM formulations on the objective of maximizing the number of on-time tasks.

The overall results with varied n, λ, c_ℓ, q_ℓ, and sf_ℓ are shown in Figure 6.6. MaxSAT encodings (both PM and WPM) solved substantially more instances than SMT within the time limit in all cases. As to the performance comparison between PM and WPM, the completion percentage of WPM is always no less than that of PM no matter how the experimental parameters change. To be specific, with the number of tasks n increasing from 50 to 300, both PM and WPM solved all the instances within the time limit (Figure 6.6a). In comparison, as any of the parameters λ, c_ℓ, q_ℓ, and sf_ℓ increases, WPM formulation gradually surpasses PM with more completed schedules (Figure 6.6b–e).

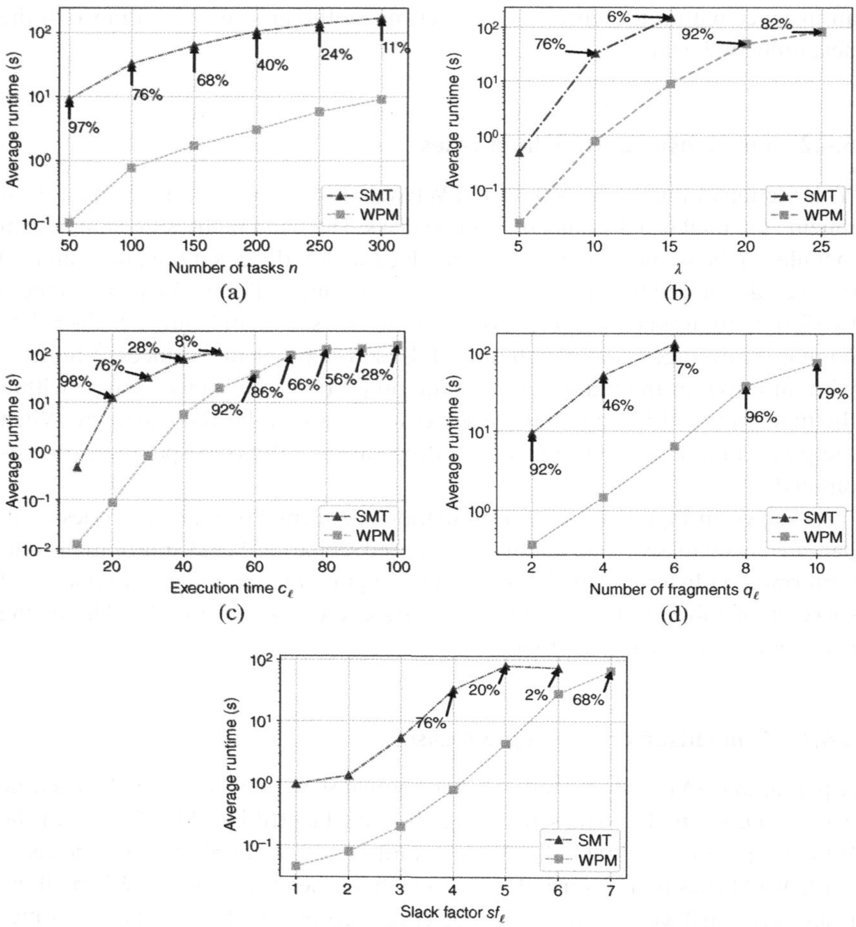

Figure 6.5 Performance comparison of WPM and SMT for weighted cases in scenarios 1–5. (a) S1: varied number of tasks n, (b) S2: varied task arriving rate λ, (c) S3: varied execution time c_ℓ, (d) S4: varied number of fragments q_ℓ, and (e) S5: varied slack factor sf_ℓ (Source: Xiaojuan Liao et al., 2021/Hindawi/CC BY 4.0).

To compare the performance of PM and WPM more clearly, we provide more criteria for evaluation. The first one is the average runtime of PM and WPM formulations on their respective completed schedules. Note that when PM and WPM solved different numbers of instances within the time limit, the average runtime on solved instances may not clearly reflect the efficiency of these two formulations. For example, when $q_\ell \sim DU(1, 12)$, WPM solved more instances

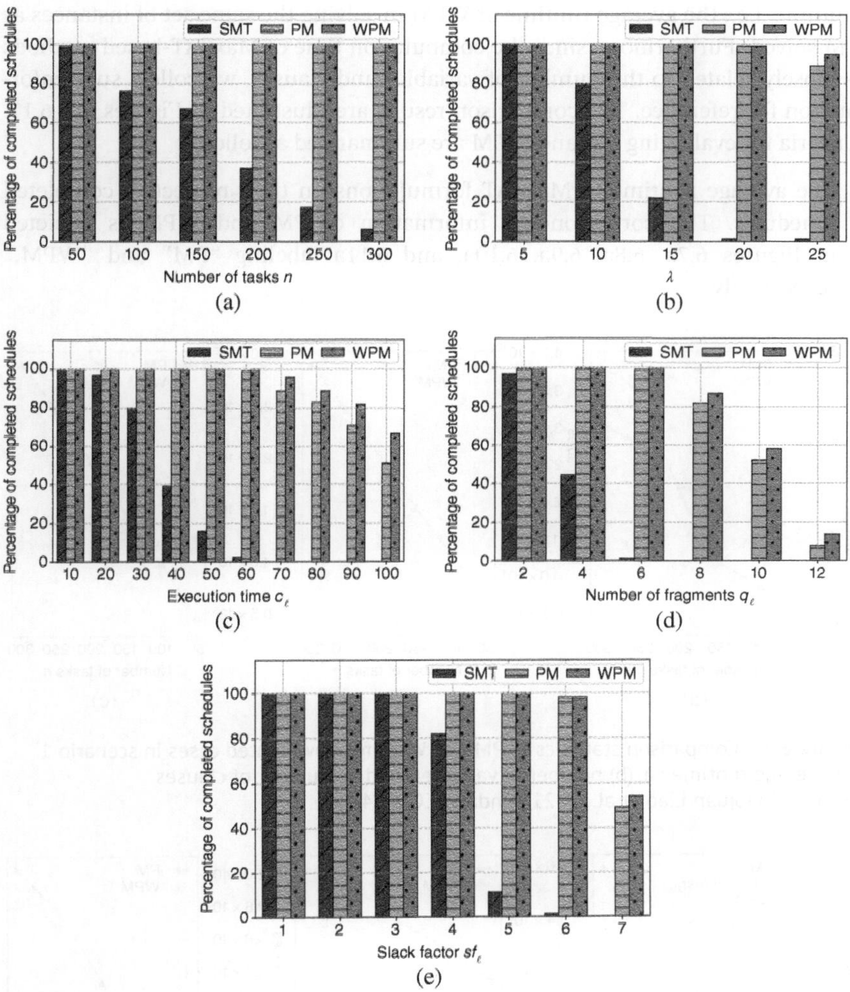

Figure 6.6 Completion percentage of formulations for unweighted cases in scenarios 1–5. (a) S1: varied number of tasks n, (b) S2: varied task arriving rate λ, (c) S3: varied execution time c_ℓ, (d) S4: varied number of fragments q_ℓ, and (e) S5: varied slack factor sf_ℓ (Source: Xiaojuan Liao et al., 2021/Hindawi/CC BY 4.0).

than PM (as shown in Figure 6.6d) and meanwhile consumed longer time on solved instances (as shown in Figure 6.10a). In this case, the average runtime is on longer the proper metrics to indicate the performance. On the other hand, we notice that in all the tested scenarios, the instances that were solved by PM could also be addressed by WPM. Therefore, we provide a secondary criterion on runtime, i.e., the average runtime of WPM in solving the same set of instances as PM solved. Furthermore, since the computation time of MaxSAT-based methods is closely related to the number of variables and clauses, we collect such information for reference. The comparison results are illustrated in Figures 6.7–6.11. Criteria for evaluating PM and WPM are summarized as follows:

- The average runtime of MaxSAT formulations on their respective completed schedules. The corresponding information of PM and WPM is depicted in Figures 6.7a, 6.8a, 6.9a, 6.10a, and 6.11a labeling "PM" and "WPM," respectively.

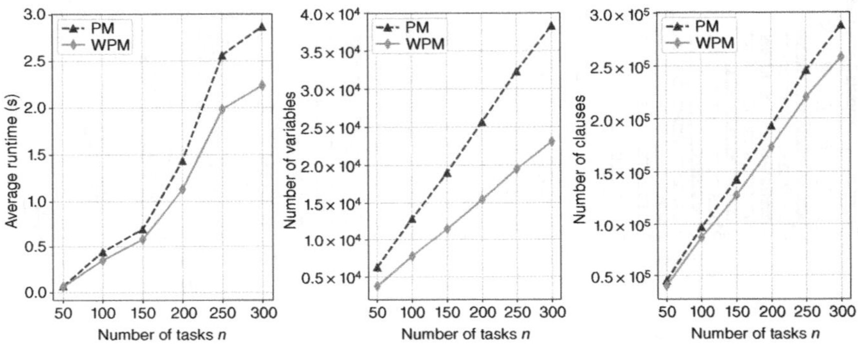

(a) (b) (c)

Figure 6.7 Comparison statistics of PM and WPM for unweighted cases in scenario 1. (a) Average runtime (s), (b) number of variables, and (c) number of clauses (Source: Xiaojuan Liao et al., 2021/Hindawi/CC BY 4.0).

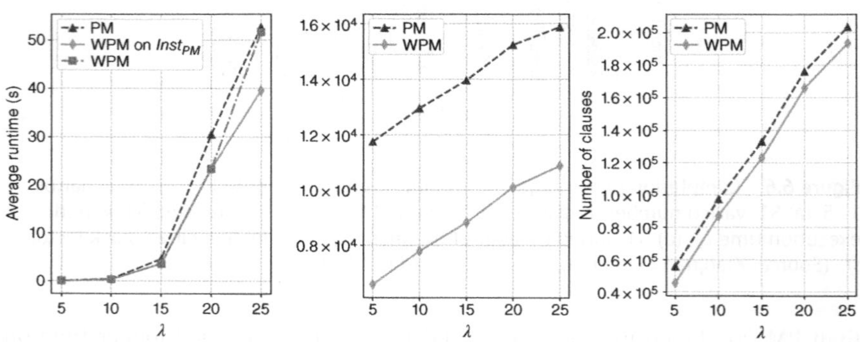

(a) (b) (c)

Figure 6.8 Comparison statistics of PM and WPM for unweighted cases in scenario 2. (a) Average runtime (s), (b) number of variables, and (c) number of clauses (Source: Xiaojuan Liao et al., 2021/Hindawi/CC BY 4.0).

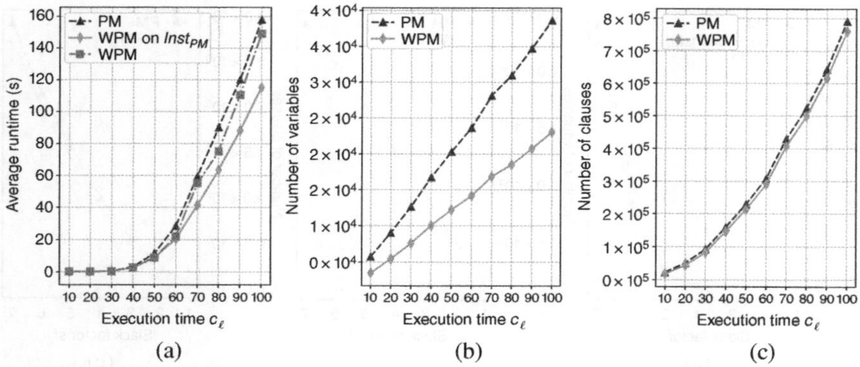

Figure 6.9 Comparison statistics of PM and WPM for unweighted cases in scenario 3. (a) Average runtime (s), (b) number of variables, and (c) number of clauses (Source: Xiaojuan Liao et al., 2021/Hindawi/CC BY 4.0).

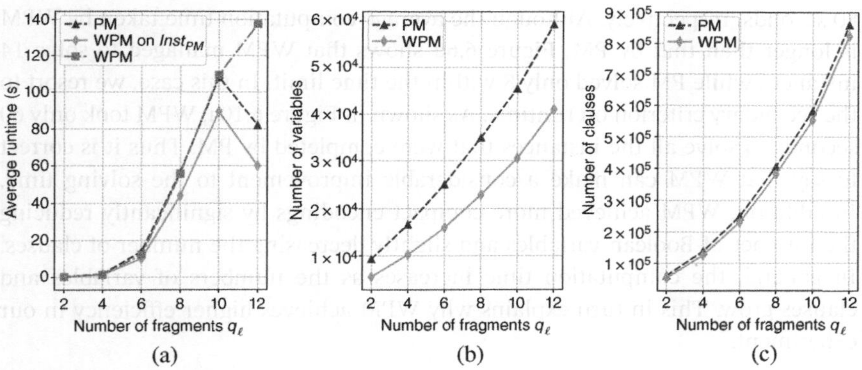

Figure 6.10 Comparison statistics of PM and WPM for unweighted cases in scenario 4. (a) Average runtime (s), (b) number of variables, and (c) number of clauses (Source: Xiaojuan Liao et al., 2021/Hindawi/CC BY 4.0).

- The average runtime of WPM in solving the same set of instances as PM solved. The corresponding information is depicted with label "WPM on $Inst_{PM}$." If the percentages of schedules completed by PM and WPM are the same, then "WPM on $Inst_{PM}$" is omitted (e.g., Figure 6.10a).
- The average number of variables generated by MaxSAT formulations.
- The average number of clauses generated by MaxSAT formulations.

Overall, the WPM formulation is more efficient than PM in both runtime criteria. The only exception appears in Figure 6.10a, where the average runtime of WPM and PM on solved instances when $q_\ell \sim DU(1, 12)$ is around 140 and

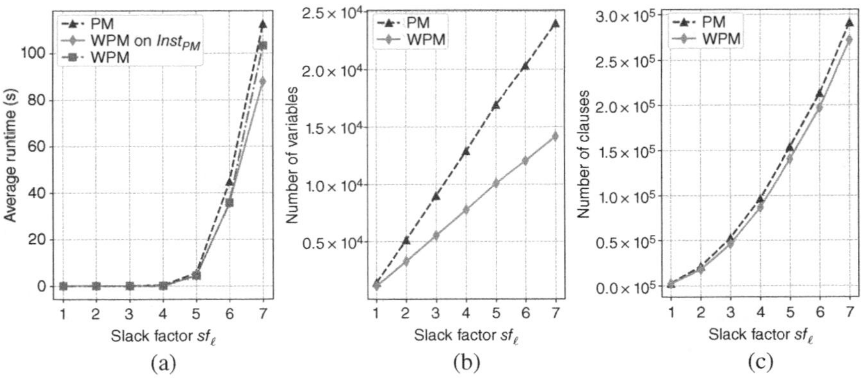

Figure 6.11 Comparison statistics of PM and WPM for unweighted cases in scenario 5. (a) Average runtime (s), (b) number of variables, and (c) number of clauses (Source: Xiaojuan Liao et al., 2021/Hindawi/CC BY 4.0).

80 seconds, respectively. Although the average computation time taken by WPM is longer than that by PM, Figure 6.6d shows that WPM managed to solve 14 instances while PM solved only 8 within the time limit. In this case, we resort to the secondary criterion on runtime. As shown in Figure 6.10a, WPM took only 60 seconds to solve all the instances that were completed by PM. Thus it is correct to say that WPM can make a considerable improvement to the solving time. In addition, WPM achieved more compact encodings by significantly reducing the number of Boolean variables and slightly decreasing the number of clauses. In general, the computation time increases as the numbers of variables and clauses grow. This in turn explains why WPM achieves higher efficiency in our experiment.

6.5 Adaption for Parallel-machine Scheduling Problem

While the WPM formulation is designed for single-machine scheduling, the encoding can be extended to parallel identical machines with minor changes. This section shows how to apply the WPM formulation to parallel identical machine scheduling problems.

The parallel-machine scheduling involves processing n tasks $\Gamma = \{\tau_1, \dots, \tau_n\}$ on m identical machines $\mathcal{M} = \{M_1, M_2, \dots, M_m\}$. Each machine can handle only one task at a time, and each task cannot be processed in parallel. Other settings of the problem are consistent with the single-machine scheduling problem described in Section 6.1. Given a set of tasks Γ and that of machines $\mathcal{M}$, the optimization problem is not only to find a schedule that gives the start execution time of each fragment $f_i^\ell \in \tau_\ell$ in Γ, but also a mapping for f_i^ℓ to a machine in $\mathcal{M}$.

To indicate on which machine a fragment is executed, $\forall \tau_\ell \in \Gamma$, $\forall f_i^\ell \in \tau_\ell$, $\forall M_u \in \mathcal{M}$, we extend the single-machine scheduling formulation by introducing one more Boolean variable $b_{i,u}^\ell$. $b_{i,u}^\ell = 1$ if f_i^ℓ is executed by M_u and $b_{i,u}^\ell = 0$ otherwise. Correspondingly, rule (C3) in Section 6.2.2 is replaced with the following two additional constraints:

(R1) $\forall \tau_\ell \in \Gamma, \forall f_i^\ell \in \tau_\ell, f_i^\ell$ is allocated and executed on a single machine:

$$\bigvee_{u=1,\dots,m} b_{i,u}^\ell \quad (1 \le \ell \le n, 1 \le i \le q_\ell) \tag{6.15}$$

(R2) $\forall \tau_k, \tau_\ell \in \Gamma$, $\forall f_i^k \in \tau_k$, $\forall f_j^\ell \in \tau_\ell$, if $k \neq l$, $\tau_k \prec \tau_\ell$, $\tau_\ell \prec \tau_k$, $EST_i^k < LCT_j^\ell$ and $EST_j^\ell < LCT_i^k$, and then f_i^k and f_j^ℓ may require the processor at the same time. In this condition, f_i^k precedes f_j^ℓ or f_j^ℓ precedes f_i^k:

$$b_{i,u}^k \wedge b_{i,u}^\ell \rightarrow pr_{i,j}^{k,\ell} \vee pr_{j,i}^{\ell,k} \tag{6.16}$$

We refer to the set of hard clauses introduced in (R1) and (R2) as $\mathcal{R}$, and rules in (C1)–(C7) excluding (C3) as $\mathcal{C}'$. Conjuncted with the set of weighted soft clauses $\mathcal{O}$ defined in Section 6.2.2, the problem is $\{\mathcal{C}', \mathcal{R}, \mathcal{O}\}$, which tries to find an assignment of variables to satisfy all the hard clauses in $\mathcal{C}' \cup \mathcal{R}$ and to maximize the sum of the weights of the satisfied soft clauses in $\mathcal{O}$. Thus, the parallel identical machine scheduling problem of maximizing the total weight of on-time tasks can be optimally solved with any off-the-shelf WPM solver.

6.6 Chapter Summary

This chapter concentrated on a Weight Partial MaxSAT (WPM) formulation for optimal scheduling in overloaded situations, with the objective of maximizing the weighted number of on-time tasks. Thanks to the WPM feature that distinguishes between hard clauses and weighted soft clauses, the task properties were naturally encoded as hard clauses and the goal of completing tasks on time were treated as a set of weighted soft clauses. Subsequently, a readily available WPM solver was employed to satisfy all the hard clauses and maximize the total weight of satisfied soft clauses. The optimal schedule can then be derived from the output of the WPM solver.

The WPM formulation's performance was compared with that of existing SMT and partial MaxSAT (PM) formulations. First, the performance of WPM was compared with SMT, demonstrating that WPM is significantly superior to SMT. Then,

WPM and PM formulations were compared in a special case with equal weights. Results indicated that the WPM formulation achieves more compact encoding and higher efficiency than PM. Furthermore, the WPM formulation was adapted for parallel identical machines with minimal modifications, underscoring its flexibility and scalability.

The restricted preemptive scheduling model considered in this chapter is built upon earlier works [9–12], where each task is split into multiple non-preemptive fragments, and preemptions can only occur at the boundaries of these fragments. However, this preemptive model with fixed points may be too inflexible for application in many practical systems. Exploring variations of preemptive models may be more promising. One example is to execute tasks continuously without interruption for at least a certain portions of time [17–20], which is the focus of Chapter 7. Instead of defining subtasks in advance, the restricted preemptive model only guarantees the minimum level of "granularity" for preemption, making it more suitable for real-world applications.

References

1 C. E. Aguero, N. Koenig, I. Chen, H. Boyer, S. Peters, J. Hsu, B. Gerkey, S. Paepcke, J. L. Rivero, and J. Manzo. Inside the virtual robotics challenge: Simulating real-time robotic disaster response. *IEEE Transactions on Automation Science and Engineering*, 12(2):494–506, 2015.

2 S. B. Er and R. E. Smith. Method and apparatus for monitoring and displaying lead impedance in real-time for an implantable medical device. US Patent 5,891,179, 1999.

3 D. P. Xenos, M. Cicciotti, G. M. Kopanos, A. E. F. Bouaswaig, O. Kahrs, R. Martinez-Botas, and N. F. Thornhill. Optimization of a network of compressors in parallel: Real time optimization (RTO) of compressors in chemical plants - an industrial case study. *Applied Energy*, 144(5):51–63, 2015.

4 G. D. Baulier, S. M. Blott, B. L. Branch, T. M. Cliff, H. F. Korth, J. E. Polito, A. Silberschatz, and S. L. Speicher. Real-time event processing system for telecommunications and other applications. US6496831B1, 2002.

5 L. Abeni, G. Buttazzo, S. Superiore, and S. Anna. Integrating multimedia applications in hard real-time systems. In *IEEE Real-time Systems Symposium*, 1998.

6 S. K. Baruah and J. R. Haritsa. Scheduling for overload in real-time systems. *IEEE Transactions on Computers*, 46(9):1034–1039, 1997.

7 S. K. Baruah, J. R. Haritsa, and N. Sharma. On-line scheduling to maximize task completions. In *Proceedings of the 15th IEEE Real-Time Systems Symposium (RTSS '94)*, San Juan, Puerto Rico, December 7–9, 1994, pages 228–236. IEEE Computer Society, 1994.

8 Z. Cheng, H. Zhang, Y. Tan, and Y. Lim. Scheduling overload for real-time systems using SMT solver. In *IEEE/ACIS International Conference on Software Engineering, Artificial Intelligence, Networking and Parallel/distributed Computing*, pages 189–194, Shanghai, China, July 2016. IEEE.

9 Z. Cheng, H. Zhang, Y. Tan, and Y. Lim. SMT-based scheduling for overloaded real-time systems. *IEICE Transactions on Information and Systems*, E100-D(5):1055–1066, 2017.

10 S. Wang, X. Liao, M. Wang, L. Chang, H. Yang, and T. Wang. An improved SMT-based scheduling for overloaded real-time systems. *Engineering Letters*, 28(1):112–122, 2020.

11 X. Liao, H. Zhang, M. Koshimura, R. Huang, and W. Yu. Maximum satisfiability formulation for optimal scheduling in overloaded real-time systems. In *PRICAI 2019: Trends in Artificial Intelligence - 16th Pacific Rim International Conference on Artificial Intelligence*, Cuvu, Yanuca Island, Fiji, August 26–30, 2019, *Proceedings, Part I*, volume 11670 of *Lecture Notes in Computer Science*, pages 618–631. Springer, 2019.

12 X. Liao, H. Zhang, M. Koshimura, R. Huang, W. Yu, and F. Li. Modeling and solving scheduling in overloaded situations with weighted partial MaxSAT. *Mathematical Problems in Engineering*, 2021:1–17, 2021.

13 J. M. Crawford and A. B. Baker. Experimental results on the application of satisfiability algorithms to scheduling problems. In *12 AAAI National Conference on Artificial Intelligence*, pages 1092–1097, 445 Burgess Drive Menlo Park, CA, United States, October 1994. American Association for Artificial Intelligence.

14 M. Koshimura, H. Nabeshima, H. Fujita, and R. Hasegawa. Solving open job-shop scheduling problems by SAT encoding. *IEICE Transactions on Information and Systems*, E93-D(8):2316–2318, 2010.

15 M. Koshimura, T. Zhang, H. Fujita, and R. Hasegawa. QMaxSAT: A partial Max-SAT solver. *Journal on Satisfiability, Boolean Modeling and Computation*, 8:95–100, 2012.

16 L. De Moura and N. Bjτrner. Z3: An efficient SMT solver. In *International Conference on Tools and Algorithms for the Construction and Analysis of Systems*, pages 337–340, Berlin, Heidelberg, March 2008. Springer-Verlag.

17 K. Ecker and R. Hirschberg. Task scheduling with restricted preemptions. In A. Bode, M. Reeve, and G. Wolf, editors, *PARLE '93 Parallel Architectures and Languages Europe*, pages 464–475, Berlin, Heidelberg, June 1993. Springer-Verlag.

18 T. Barański. Task scheduling with restricted preemptions. In *2011 Federated Conference on Computer Science and Information Systems (FedCSIS)*, pages 231–238, Szczecin, Poland, November 2011. IEEE.

19 K. Pieńkosz and A. Prus. Task scheduling with restricted preemptions on two parallel processors. In *2015 20th International Conference on Methods and Models in Automation and Robotics (MMAR)*, pages 58–61, Miedzyzdroje, Poland, August 2015. IEEE.

20 X. Liao, H. Zhang, M. Koshimura, R. Huang, and F. Li. Solving restricted preemptive scheduling on parallel machines with SAT and PMS. *Journal of Universal Computer Science*, 29(8):911–937, 2023.

7

Restricted Preemptive Scheduling

After reading this chapter, you should be able to:

- Understand the model of k-restricted preemptive scheduling on parallel machines.
- Grasp the process of encoding k-restricted preemptive scheduling into SAT and partial MaxSAT (PMS) problems.
- Explain the phenomena observed in experiments and provide a rational comparison between the SAT and PMS approaches.
- Comprehend the way of evaluating heuristic algorithms.

7.1 Restricted Preemptive Scheduling Model

This chapter addresses the k-restricted preemptive scheduling problem on parallel identical machines, where preemptions are only allowed after a task has been continuously executed for at least k units of time. The objective is to minimize the makespan C, which is the maximum completion time on all the machines. This deterministic scheduling problem is represented in three-field notation as $P|k-pmtn|C_{max}$ [1, 2].

In the $P|k-pmtn|C_{max}$ problem, n tasks $\mathscr{T} = \{T_1, T_2, \dots, T_n\}$ must be processed on m identical machines[1] $\mathscr{M} = \{M_1, M_2, \dots, M_m\}$. Each machine can handle only one task at a time, and no task can be executed in parallel. Processing time of task $T_j \in \mathscr{T}$ is given by p_j, known in advance. Preemptions are restricted by granularity $k \in \mathbb{N}$, meeting the following condition: $\forall T_j \in \mathscr{T}$, if p_j is less than or equal to k, then preemption is prohibited, otherwise preemption may take place after T_j has been continuously processed for at least k units of time.

1 The term "machine" can refer to any actual or abstract machine or processor that executes tasks. In this chapter, we do not distinguish "machines" and "processors."

Applied Satisfiability: Cryptography, Scheduling, and Coalitional Games, First Edition. Xiaojuan Liao and Miyuki Koshimura.

Table 7.1 Research status of $P|k-pmtn|C_{max}$ problem.

Conditions	Complexity	Solution	Literature
General cases	NP-hard	Optimal[a)]	Ecker and Hirschberg [2]
General cases	$O(n \cdot \log(n))$	Suboptimal	Barański [6]
Small-scaled general cases	NP-hard	Optimal	Liao et al. [5]
General cases with $m = 2$	Open	Optimal	Knust et al. [4]
General cases with $m = 2$	$O(n^2)$	Suboptimal	Pieńkosz and Prus [3]
General cases with $m = 2$	$O(n \cdot \log(n))$	Suboptimal	Pieńkosz and Prus [3]
$m = 2$ and k is fixed.	$O(n)$	Optimal	Ecker and Hirschberg [2]
$m = 2, p_1 \geq \frac{1}{2}\sum_{T_j \in \mathcal{T}} p_j$	$O(n)$	Optimal	Ecker and Hirschberg [2]
$m = 2, p_1 \geq 4k$	$O(n)$	Optimal	Pieńkosz and Prus [3]
$m = 2, p_1 \geq 2k, p_2 \geq 3k$	$O(n)$	Optimal	Pieńkosz and Prus [3]
$m = 2, p_j \geq 2k$	$O(n)$	Optimal	Knust et al. [4]

a) No optimal algorithm is given.

Ecker and Hirschberg [2] demonstrated that the $P|k-pmtn|C_{max}$ problem is NP-hard, with some exceptions for special cases on two identical machines. For example, Pieńkosz and Prus [3] showed that the two-machine problem can be solved in $O(n)$ time if one task has a processing time of at least $4k$, or if one task has a processing time of at least $2k$ and another has a processing time of at least $3k$. Knust et al. [4][2] presented an $O(n)$ algorithm for two identical machines, assuming that all tasks have processing times of at least $2k$.

Table 7.1 summarizes the current research on this problem. As far as we know, Liao et al. [5] is the only study that has proposed optimal algorithms for addressing the $P|k-pmtn|C_{max}$ problem. This chapter introduces a mathematical programming (MP) model and the work of Liao et al. [5], which presents two satisfiability-based approaches for tackling the small-scaled $P|k-pmtn|C_{max}$ problem.

In the $P|k-pmtn|C_{max}$ problem, preemption is allowed after a task has been processed for at least k units of time. That is, each task $T_j \in \mathcal{T}$ can be divided into $\ell_j = \lceil p_j/k \rceil$ indivisible subtasks $\langle T_j^1, \ldots, T_j^{\ell_j} \rangle$. The subtasks must meet two conditions. First, for each $T_j \in \mathcal{T}$ and $T_j^i \in T_j \backslash \{T_j^{\ell_j}\}$, the processing time p_j^i must be at least k or zero. If $p_j^i = 0$, T_j^i is *void* and cannot be assigned to any machine. If $p_j^i > 0$, T_j^i is *valid* and must be processed by one machine. Second, the total processing time

2 Knust et al. [4] showed that the algorithm presented for problems with pliability also applies for those with k-restricted preemptions in special cases.

of all subtasks in T_j must equal p_j, i.e., $\sum_{T_j^i \in T_j} p_j^i = p_j$. To simplify the problem at hand, several predefined constraints should be satisfied as follows:

- Tasks are independent and available at time 0.
- Each machine can handle at most one subtask at a time.
- Subtasks of each task must be processed sequentially.
- Processing times of tasks are integers.

Given a set of tasks $\mathcal{T}$ and a set of identical machines $\mathcal{M}$, solving the scheduling problem is equivalent to identifying the following critical information:

- The start execution time s_j^i of each subtask $T_j^i \in T_j$ in $\mathcal{T}$,
- The processing time p_j^i of each subtask $T_j^i \in T_j$ in $\mathcal{T}$,
- The assignment of each valid subtask $T_j^i \in T_j$ in $\mathcal{T}$ to a machine in $\mathcal{M}$.

The goal is to minimize the makespan C, defined as the time difference between the start of the earliest task and the completion of the latest task. Assuming the earliest task starts at time 0, the makespan C is $C = max\left\{\left(s_j^{\ell_j} + p_j^{\ell_j}\right) \mid T_j \in \mathcal{T}\right\}$.

The optimal makespan is represented as $C^* = \text{Min}\left(max\left\{\left(s_j^{\ell_j} + p_j^{\ell_j}\right) \mid T_j \in \mathcal{T}\right\}\right)$. C^* can be estimated as follows.

When the preemption granularity k is equal to 0, the k-restricted preemptive scheduling is equivalent to fully preemptive scheduling, which can be solved in polynomial time using McNaughton's wrap-around rule [7]. The optimal makespan for $P|0 - pmtn|C_{max}$ is

$$C^* = max\left\{p_{max}, \sum_{T_j \in \mathcal{T}} p_j/m\right\} \tag{7.1}$$

where p_{max} is the longest processing time of all the tasks, and $\sum_{T_j \in \mathcal{T}} p_j/m$ represents the average machine load. Formula (7.1) is typically used to compute the lower bound of C^* for any $k > 0$ [3, 6], represented by C_{lb}.

The upper bound of the makespan C, denoted as C_{ub}, can be estimated using heuristics such as longest processing time first (LPT) and LPT-modified algorithms [2]. The LPT algorithm assigns tasks in a nonpreemptive order to the first available machine based on the longest processing time, while the LPT-modified algorithm allocates a part of T_j of length k to the first available machine. Therefore, the optimal makespan C^* falls within the range $[C_{lb}, C_{ub}]$.

Example 7.1 Consider a task set $\mathcal{T} = \{T_1, T_2, T_3, T_4\}$, where each task has a processing time of 3, i.e., $\forall T_j \in \mathcal{T}, p_j = 3$. Suppose that the preemption granularity k is 2, meaning that each task must be executed continuously for at least 2 units of time before being preempted.

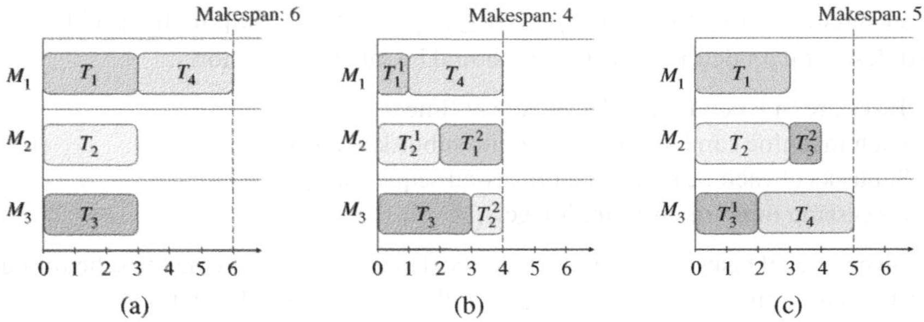

Figure 7.1 Example schedule of no preemption, fully arbitrary preemption and 2-restricted preemption. (a) Without preemption, (b) arbitrary preemption, and (c) 2-restricted preemption (Source: Xiaojuan Liao et al. (2023)/with permission of Journal of Universal Computer Science).

Figure 7.1 illustrates the optimal results for three types of scheduling on three identical machines: nonpreemptive, fully preemptive, and two-restricted preemptive scheduling. Key information is summarized as follows:

- Figure 7.1a shows the schedule generated by executing tasks without preemption, producing a makespan of 6.
- Figure 7.1b displays an average machine load with no constraints on preemptions, resulting in a makespan of 4. Note that this schedule is not feasible for two-restricted preemption because task T_1 is preempted after just one time unit.
- Figure 7.1c exhibits a feasible two-restricted schedule with a makespan equal to 5, which is also the shortest makespan achievable for the two-restricted preemptive scheduling problem.

In the two-restricted preemptive example, each task is divided into $\lceil 3/2 \rceil = 2$ subtasks. Therefore, $\forall T_j \in \mathcal{T}$, $T_j = \langle T_j^1, T_j^2 \rangle$. According to the optimal schedule shown in Figure 7.1c, the start execution time (StartTime), processing time (ProcessTime), and mapping from subtask to machine (machine identification [MID]) are summarized in Table 7.2. Subtasks T_1^2, T_2^2, and T_4^2 are void, thus the their start times and machine identifiers are marked by slashes.

7.2 Mathematical Programming

This section explains how to model the $P|k\text{-pmtn}|C_{max}$ problem using MP. This approach lays the foundation for understanding subsequent satisfiable (SAT) methods and serves as a benchmark for evaluating satisfiability-based methods.

Table 7.2 Critical information of two-restricted preemptive scheduling in Figure 7.1c.

	Subtask 1			Subtask 2		
	StartTime	**ProcessTime**	**MID**	**StartTime**	**ProcessTime**	**MID**
T_1	0	3	1	/	0	/
T_2	0	3	2	/	0	/
T_3	0	2	3	3	1	2
T_4	2	3	3	/	0	/

Objective:	$Min(\max_{\forall T_j \in \Gamma}(s_j^{\ell_j} + p_j^{\ell_j}))$	
st:		
(C1)	$\forall T_j \in \Gamma, \forall T_j^i, T_j^{i+1} \in T_j$	$s_j^{i+1} \geq s_j^i + p_j^i$
(C2)	$\forall T_i, T_j \in \Gamma, \forall T_i^x \in T_i, \forall T_j^y \in T_j, \forall M_u \in \mathcal{M}$	$b_i^{x,u} \wedge b_j^{y,u} \Rightarrow \left(s_i^x \geq s_j^y + p_j^y\right) \vee \left(s_j^y \geq s_i^x + p_i^x\right)$
(C3)	$\forall T_j \in \Gamma, \forall T_j^i \in T_j \backslash \{T_j^{\ell_j}\}$	$(p_j^i \geq k) \vee (p_j^i = 0)$
(C4)	$\forall T_j \in \Gamma, \forall T_j^i \in T_j$	$(p_j^i = 0) \Rightarrow \sum_{M_u \in \mathcal{M}} b_j^{i,u} = 0$
(C5)	$\forall T_j \in \Gamma, \forall T_j^i \in T_j$	$(p_j^i \geq \epsilon) \Rightarrow \sum_{M_u \in \mathcal{M}} b_j^{i,u} = 1$
(C6)	$\forall T_j \in \Gamma, \forall T_j^i \in T_j$	$\sum_{T_j^i \in T_j} p_j^i = p_j$
(C7)	$\forall T_j \in \Gamma, \forall T_j^i \in T_j$	$0 \leq s_j^i \leq L_{\max}, 0 \leq p_j^i \leq p_j, s_j^i, p_j^i \in \mathbb{Z}$
(C8)	$\forall T_j \in \Gamma, \forall T_j^i \in T_j, \forall M_u \in \mathcal{M}$	$b_j^{i,u} \in \{0, 1\}$

Figure 7.2 Mathematical programming of $P|k - pmtn|C_{max}$.

First, 0–1 decision variables $b_j^{i,u}$ are introduced to describe on which machine a subtask is executed. $\forall T_j \in \Gamma, \forall T_j^i \in T_j, \forall M_u \in \mathcal{M}$:

$$b_j^{i,u} = \begin{cases} 1, & \text{if } T_j^i \text{ is executed by } M_u \\ 0, & \text{otherwise} \end{cases} \tag{7.2}$$

Then, the MP formulation of the problem $P|k - pmtn|C_{max}$ is defined, shown in Figure 7.2. Symbol $\Longrightarrow$ is the logical implication operator and ϵ is a small positive real number approaching zero.

Constraint (C1) guarantees that all the portions of the same task are executed in sequence. Constraint (C2) ensures that if two portions from different tasks are assigned to the same machine, then one should precede the other. Constraint (C3) states that the processing time of all subtasks is either no less than k or equal to 0, except for the last portion of a task. Given a subtask $T_j^i \in T_j$ in Γ, constraints (C4) and (C5) guarantee that if $p_j^i = 0$, then the void subtask T_j^i cannot be processed by any machines, and on the other hand, if $p_j^i > 0$, then valid subtask T_j^i should be allocated and executed on a single machine. Constraint (C6) ensures that the sum of processing time of all subtasks in T_j is equal to p_j. Finally, constraints (C7) and (C8) restricts the ranges of variables in the formulation. The objective of minimizing the makespan is given by $Min(max_{\forall T_j \in \Gamma \forall T_j^i \in T_j}(s_j^{\ell} + p_j^{\ell}))$, under the condition that constraints (C1)–(C7) are all satisfied.

7.3 SAT Approach

This section outlines the conversion of the $P|k - pmtn|C_{max}$ problem into a SAT problem. To encompass all critical constraints of the scheduling model, we introduce five types of Boolean variables:

- **Task allocation variables**: $b_j^{i,u}$ $(1 \leq j \leq n, 1 \leq i \leq \ell_j, 1 \leq u \leq m)$, which is true if T_j^i is assigned to machine M_u.
- **Schedule digit variables**: $sd_j^{i,v}$ $(1 \leq j \leq n, 1 \leq i \leq \ell_j, 0 \leq v \leq \lfloor \log_2 C \rfloor)$ for each s_j^i, where C is the schedule length within which all tasks should be completed. $sd_j^{i,v}$ is true if the vth digit of s_j^i is 1 in the binary number; otherwise, the vth digit of s_j^i is 0. That is, $s_j^i = \sum_{v=0}^{\lfloor \log_2 C \rfloor} 2^v \cdot sd_j^{i,v}$.
- **Processing time digit variables**: $pd_j^{i,v}$ $(1 \leq j \leq n, 1 \leq i \leq \ell_j, 0 \leq v \leq \lfloor \log_2 p_j \rfloor)$ for each p_j^i, which is true if the vth digit of p_j^i is 1 in the binary number, otherwise, the vth digit of p_j^i is 0. That is, $p_j^i = \sum_{v=0}^{\lfloor \log_2 p_j \rfloor} 2^v \cdot pd_j^{i,v}$.
- **Valid subtask indicator**: pk_j^i $(1 \leq j \leq n, 1 \leq i \leq \ell_j)$, which is true if $p_j^i > 0$, otherwise, $p_j^i = 0$.
- **Precedence relationship variables**: $pr_{i,j}^{x,y}$ $(1 \leq i \leq n-1, i+1 \leq j \leq n, 1 \leq x \leq \ell_i, 1 \leq y \leq \ell_j)$, which is introduced for the case that T_i^x and T_j^y are executed by the same machine u. $pr_{i,j}^{x,y} = 1$ if T_i^x precedes T_j^y when $b_i^{x,u} \wedge b_j^{y,u}$, otherwise, T_j^y precedes T_i^x.

Using these Boolean variables, we create rules to encode the $P|k$-pmtn$|C_{max}$ problem into a set of clauses. *PB(Exp)* is a collection of clauses derived from SAT encoding of the linear pseudo-Boolean expression *Exp* of the form $\sum_j a_j l_j \{\geq, \leq, =\} b$, where a_j and b are integers, and l_j are literals. Pseudo-Boolean constraints are significant in various practical applications, and numerous methods exist to convert pseudo-Boolean formulations to conjunctive normal form (CNF) formulas [8], such as binary decision diagram (BDD), sequential weight counters, sorting networks, adder networks, and binary merge. These encodings are all integrated into the Python toolkit PySAT [9].

(S1) Subtask order constraint, which ensures subtasks within each task are executed in order: $\forall T_j \in \mathcal{T}, \forall T_j^i, T_j^{i+1} \in T_j$, T_j^i precedes T_j^{i+1}.

$$PB\left(\sum_{v=0}^{\lfloor \log_2 C \rfloor} 2^v \cdot sd_j^{i+1,v} \geq \sum_{v=0}^{\lfloor \log_2 C \rfloor} 2^v \cdot sd_j^{i,v} + \sum_{v=0}^{\lfloor \log_2 p_j \rfloor} 2^v \cdot pd_j^{i,v}\right) \tag{7.3}$$

(S2) Precedence relationship constraint, which ensures correct order between subtasks of different tasks assigned to the same machine: $\forall T_i, T_j \in \mathcal{T}$ $(i < j)$, $\forall T_i^x \in T_i, \forall T_j^y \in T_j, \forall M_u \in \mathcal{M}$, if T_i^x and T_j^y are assigned to the same machine, then T_i^x precedes T_j^y or T_j^y precedes T_i^x.

$$\begin{aligned} &\neg b_i^{x,u} \vee \neg b_j^{y,u} \vee pr_{i,j}^{x,y} \vee C \quad (\forall C \in P_1) \\ &\neg b_i^{x,u} \vee \neg b_j^{y,u} \vee \neg pr_{i,j}^{x,y} \vee C \quad (\forall C \in P_2) \end{aligned} \tag{7.4}$$

where

$$\begin{aligned} P_1 &= PB\left(\sum_{v=0}^{\lfloor \log_2 C \rfloor} 2^v \cdot sd_j^{y,v} \geq \sum_{v=0}^{\lfloor \log_2 C \rfloor} 2^v \cdot sd_i^{x,v} + \sum_{v=0}^{\lfloor \log_2 p_i \rfloor} 2^v \cdot pd_i^{x,v}\right) \\ P_2 &= PB\left(\sum_{v=0}^{\lfloor \log_2 C \rfloor} 2^v \cdot sd_i^{x,v} \geq \sum_{v=0}^{\lfloor \log_2 C \rfloor} 2^v \cdot sd_j^{y,v} + \sum_{v=0}^{\lfloor \log_2 p_j \rfloor} 2^v \cdot pd_j^{y,v}\right) \end{aligned} \tag{7.5}$$

(S3) Valid subtask allocation, which ensures valid subtasks are allocated to machines: $\forall T_j \in \mathcal{T}, \forall T_j^i \in T_j$, if T_j^i is valid, i.e., $p_j^i > 0$, then T_j^i is allocated and executed on a single machine.

$$\neg pk_j^i \vee \bigvee_{M_u \in \mathcal{M}} b_j^{i,u} \tag{7.6}$$

Furthermore, if a valid subtask T_j^i is not the last subtask of T_j, i.e., $T_j^i \in T_j \backslash \{T_j^{\ell_j}\}$, then $p_j^i \geq k$:

$$\neg pk_j^i \vee C \quad \left(\forall C \in PB\left(\sum_{v=0}^{\lfloor \log_2 p_j \rfloor} 2^v \cdot pd_j^{i,v} \geq k\right)\right) \tag{7.7}$$

(S4) Void subtask constraint, which ensures void subtasks are not allocated to machines: $\forall T_j \in \mathcal{T}, \forall T_j^i \in T_j$, if T_j^i is void, i.e., $p_j^i = 0$, then T_j^i cannot be allocated on any machines.[3]

$$pk_j^i \vee \neg b_j^{i,u} \quad (\forall M_u \in \mathcal{M}) \tag{7.8}$$

---

3 Assigning void subtasks to machines is possible, but this would increase computation time especially for small k. The reason is explained as follows. That k is small means that tasks can be split into a large number of small pieces, thus much effort has to be made to allocate these (valid and void) subtasks. The rule (S4) prevents void subtasks from being executed, thus the solver does not bother to consider the allocation of these subtasks.

Furthermore, $\forall v \in \{0, 1, \dots, \lfloor \log_2 p_j \rfloor\}$, $pd_j^{i,v} = 0$ for any void subtask T_j^i:

$$pk_j^i \vee \neg pd_j^{i,v} \tag{7.9}$$

(S5) Individual processing time constraint: $\forall T_j \in \mathcal{T}$, the sum of processing time of all subtasks in T_j is equal to p_j.

$$PB\left(\sum_{T_j^i \in T_j} \sum_{v=0}^{\lfloor \log_2 p_j \rfloor} 2^v \cdot pd_j^{i,v} = p_j\right) \tag{7.10}$$

(S6) Task completion time constraint: $\forall T_j \in \mathcal{T}$, T_j should end by C.

$$PB\left(\sum_{v=0}^{\lfloor \log_2 C \rfloor} 2^v \cdot sd_j^{\ell,v} + \sum_{v=0}^{\lfloor \log_2 p_j \rfloor} 2^v \cdot pd_j^{\ell,v} \le C\right) \tag{7.11}$$

At this stage, the $P|k-pmtn|C_{max}$ problem has been transformed into Boolean formulas, which are then converted into a set of clauses. These clauses are combined using logical *and* to create a Boolean propositional formula in CNF. An off-the-shelf SAT solver is used to check the satisfiability of this formula. We start by setting the initial makespan value C to C_{lb}. If the SAT solver finds a nonempty solution model, it means a valid variable assignment has been found that satisfies the formula, thus confirming the optimal makespan is C_{lb}. If the solver returns an empty model, it indicates that C_{lb} is too short to complete all tasks. In this scenario, we adjust the value of C by performing a binary search within the interval $[C_{lb}+1, C_{ub}]$ to determine the optimal makespan C^*. Algorithm 7.1 outlines the procedure for determining the optimal makespan for the $P|k-pmtn|C_{max}$ problem using SAT formulation.

The input is the $P|k-pmtn|C_{max}$ problem, represented as $\mathcal{T}, m, k$, where $\mathcal{T}$ is a set of tasks with specified processing times, $m = |\mathcal{M}|$ is the number of identical machines, and k is the preemption granularity. The output includes the optimal schedule **S** and mapping **M**. Schedule **S** specifies the exact start time and processing duration for each subtask in $T_j \in \mathcal{T}$, while **M** maps each valid subtask in $T_j \in \mathcal{T}$ to a machine in $\mathcal{M}$.

The function *Encoding*($\mathcal{T}, m, k, C$) produces a set of clauses based on constraints (S1)–(S6). The function *CallSATSolver*($\mathcal{C}$) invokes a SAT solver to find a solution model $\mathcal{D}$ that satisfies all clauses in $\mathcal{C}$. If such a model exists, *CallSATSolver*($\mathcal{C}$) returns it; otherwise, it returns an empty model $\emptyset$.

Initially, the problem $\{\mathcal{T}, m, k, C_{lb}\}$ is encoded into a set of clauses $\mathcal{C}$, assuming all tasks can be completed within the minimum makespan C_{lb} (line 1). A SAT solver checks for the existence of a solution model that satisfies $\mathcal{C}$ (line 2). If a nonempty model $\mathcal{D}$ is found (line 3), the optimal makespan is confirmed to

Algorithm 7.1 Optimal makespan identification by SAT formulation

Input: A k-restricted preemptive scheduling problem $\{\mathcal{T}, m, k\}$. $\mathcal{T}$ is a set of tasks, $m = |\mathcal{M}|$ is the number of identical machines, and k is the preemption granularity.

Output: Schedule **S** and mapping **M**.

1: $C = Encoding(\mathcal{T}, m, k, C_{lb})$
2: $D = CallSATSolver(C)$
3: **if** $D \neq \emptyset$ **then**
4:     return $\{\mathbf{S}, \mathbf{M}\}$ based on model D ▷ C_{lb} is the optimal makespan.
5: **end if**
6: $left \leftarrow C_{lb} + 1$; $right \leftarrow C_{ub}$ ▷ Perform binary search for satisfaction between $[C_{lb} + 1, C_{ub}]$.
7: **while** $left < right$ **do**
8:     $mid \leftarrow \lfloor left + right \rfloor / 2$
9:     $C = Encoding(\mathcal{T}, m, k, mid)$
10:     $D = CallSATSolver(C)$
11:     **if** $D \neq \emptyset$ **then**
12:         $right = mid$
13:     **else**
14:         $left = mid + 1$
15:     **end if**
16: **end while**
17: $C^* \leftarrow right$
18: return $\{\mathbf{S}, \mathbf{M}\}$ based on model D

be C_{lb}, and the corresponding schedule **S** and mapping **M** are derived from $\mathscr{D}$ (line 4).

Lines 6–18 address the scenario, where $C^* > C_{lb}$. The initial bounds for the optimal makespan, *left* and *right*, are set to $C_{lb} + 1$ and C_{ub}, respectively (line 6). A binary search is then conducted within the interval [*left*, *right*]. For $mid = \lfloor (left + right)/2 \rfloor$, we check if a nonempty model can be returned with the makespan *mid* (lines 8–10). If a nonempty solution model is returned (line 11), the search space is reduced by setting the upper bound to *mid* (line 12). If the model is empty (line 13), indicating the makespan is too short, the lower bound is increased to $mid + 1$ (line 14). This process continues until the search space is exhausted (line 7). The optimal makespan C^* is found to be equal to *right* (line 17), satisfying both conditions:

1) C^* makes the SAT solver return SAT.
2) $C^* - 1$ makes the SAT solver return unsatisfiable (UNSAT).

7.4 MaxSAT Approach

To eliminate the repeated calls of the SAT solver when searching for an optimal makespan greater than C_{lb}, we introduce a partial MaxSAT (PMS) formulation in this section.

The inherent limitation of SAT solvers lies in their binary decision nature: they determine whether a propositional Boolean formula is satisfiable and provide no additional information when it is unsatisfiable. As an optimized version of SAT, PMS offers an improved approach by distinguishing between hard and soft clauses within a propositional Boolean formula. Solving PMS involves finding an optimal assignment that satisfies all hard clauses and the maximum number of soft clauses. To formulate $P|k-pmtn|C_{max}$ problem with PMS, we introduce three types of Boolean variables:

- **Task allocation variables**: $x_{i,j,t}$ $(1 \le i \le m, 1 \le j \le n, 0 \le t < C_{ub})$, which is true if machine M_i handles task T_j from time t to $t+1$.
- **Individual completion time variables**: $y_{j,t}$ $(1 \le j \le n, p_j \le t \le C_{ub})$, which is true if task T_j finished by time t.
- **Global completion time variables**: z_t $(C_{lb} \le t \le C_{ub})$, which is true if all tasks in $\mathcal{T}$ finish by time t.

In what follows, we define a set of hard and soft clauses to effectively capture the scheduling constraints. Rules (H1)–(H7) generate hard clauses and (S) generates a set of soft clauses. *CNF*(*Expr*) represents a CNF formula encoded from cardinality constraint *Expr*. A cardinality constraint is a special type of pseudo-Boolean constraint that limits the number of literals evaluating true from a given set of literals [8].

(H1) Machine constraint, which ensures each machine handles at most one task at a time:

$$CNF\left(\sum_{j=1}^{n} x_{i,j,t} \le 1\right) \quad (1 \le i \le m, 0 \le t < C_{ub}) \tag{7.12}$$

(H2) Task constraint, which ensures each task is processed by at most one machine at a time:

$$CNF\left(\sum_{i=1}^{m} x_{i,j,t} \le 1\right) \quad (1 \le j \le n, 0 \le t < C_{ub}) \tag{7.13}$$

(H3) Processing time constraint, which specifies that each task T_j requires p_j time units to complete:

$$CNF\left(\sum_{i=1}^{m} \sum_{t=0}^{C_{ub}-1} x_{i,j,t} = p_j\right) \quad (1 \le j \le n) \tag{7.14}$$

(H4) Preemption constraint, which ensures tasks are processed continuously for at least k units unless completed:

$$\begin{aligned} &\neg x_{i,j,0} \vee y_{j,p} \quad \text{if } p = p_j, \\ &\neg x_{i,j,0} \vee x_{i,j,p} \quad \text{otherwise.} \end{aligned} \tag{7.15}$$

$$\begin{aligned} &x_{i,j,t} \vee \neg x_{i,j,t+1} \vee x_{i,j,(t+1)+p} \vee y_{j,(t+1)+p} \quad \text{if } t+p+1 \geq p_j, \\ &x_{i,j,t} \vee \neg x_{i,j,t+1} \vee x_{i,j,(t+1)+p} \quad \text{otherwise.} \\ &(1 \leq i \leq m, 1 \leq j \leq n) \\ &(0 \leq t < C_{ub} - p - 1, 0 < p < min\{k, p_j + 1\}) \end{aligned} \tag{7.16}$$

(H5) Task completion order, which ensures if T_j finishes by t, it finishes by $t + 1$:

$$\neg y_{j,t} \vee y_{j,t+1} \quad (1 \leq j \leq n, p_j \leq t < C_{ub}) \tag{7.17}$$

(H6) Task start time constraint, which ensures if T_j finishes by time t, it must start before time t:

$$\neg y_{j,t} \vee \neg x_{i,j,t} \quad (1 \leq i \leq m, 1 \leq j \leq n, p_j \leq t < C_{ub}) \tag{7.18}$$

(H7) Global completion constraint, which ensures z_t is true if all tasks are completed by time t.

$$\begin{aligned} &\neg z_t \vee y_{j,t} \\ &(C_{lb} \leq t \leq C_{ub}, 1 \leq j \leq n) \end{aligned} \tag{7.19}$$

(S) Makespan objective, which describes the desired completion time z_t for all tasks by time t:

$$z_t \quad (C_{lb} \leq t \leq C_{ub}) \tag{7.20}$$

The PMS problem is formulated such that all hard clauses (H1)–(H7) must be satisfied, while the soft clause (S) is optimized to minimize z_t. Assume the number of satisfied soft clauses to be n_s, then the optimal makespan is $C^* = C_{ub} - n_s + 1$. The corresponding schedule and mapping can be derived from the assignment of Boolean variables $x_{i,j,t}$ $(1 \leq i \leq m, 1 \leq j \leq n, 0 \leq t < C_{ub})$. This formulation leverages PMS's ability to handle optimization goals (soft clauses) alongside mandatory constraints (hard clauses).

7.5 Performance Evaluation

This section investigates the optimal makespan and evaluates the effectiveness of MP, SAT, and PMS approaches. The solvers chosen for these formulations are Gurobi (version 9.1.0), Glucose (version 3.0), and LSUPlus, respectively.

LLC Gurobi Optimization [10] is a robust MP solver for handling large-scale optimization problems efficiently. It can be accessed via the Python toolkit gurobipy. Glucose [11] is an incremental SAT solver built upon Minisat [12], and LSUPlus is an extension of linear search SAT-UNSAT algorithm (LSU) [13] designed specifically for MaxSAT problems. It incorporates native support for cardinality constraints, facilitated by the MiniCard solver [14]. Unlike most encodings that represent constraints like $\sum_{i=1}^{n} x_i \leq k$ as multiple clauses, the native encoding produces only a single clause, which is particularly advantageous for handling the cardinality constraints described in Section 7.4. Both Glucose and LSUPlus are integrated in Python toolkit PySAT [9]. All the experiments were carried out on the Ubuntu-20.04 virtual platform, equipped with a 3.0-GHz Intel i7-9700 processor and 16-GB RAM.

7.5.1 Evaluation on the Optimal Makespan

This section investigates the optimal makespan of the $P|k-pmtn|C_{max}$ problem. The number of tasks n is set to 10, and each task $T_j \in \mathcal{T}$ has a randomly chosen processing time p_j from the uniform distribution between 10 and 20. The preemption granularity k ranges from 5 to 19 in increments of 2, and the number of machines m ranges from 2 to 10. The reason why k starts from 5 is that heuristics [3, 6] are known to effectively find optimal solutions when k is less than half of the processing time. Figure 7.3 presents the results, where each data point represents the average makespan across 10 instances. The lower bound is calculated using McNaughton's wrap-around rule with arbitrary preemption, while the upper bound is estimated using two heuristics: the LPT algorithm without preemption and the modified LPT algorithm allowing preemption at subtask boundaries of length k.

In Figure 7.3, it is evident that for a fixed k, the optimal makespan C^* decreases as the number of machines m increases. This is expected because more machines enable tasks to be processed in parallel, thereby reducing the overall completion time.

As m increases, tasks need to be divided into smaller pieces to achieve $C^* = C_{lb}$. For instance, when $m \leq 6$, keeping k no larger than 9 ensures an even machine load that meets $C^* = C_{lb}$. However, as m reaches 7, the upper limit of k decreases to 7 for $C^* = C_{lb}$, and further reduces to 5 when $m \geq 8$.

The case of $m = 10$ (equal to the number of tasks n) is trivial since each task can be scheduled on a separate machine without preemption, resulting in overlapping lower and upper bounds with the optimal makespan. In the rest of this section, we exclude the special case where $m = n$.

For fixed m within $2 \leq m \leq 9$, the optimal makespan generally increases as k goes up, deviating from C_{lb}. This trend occurs because a smaller k enables finer

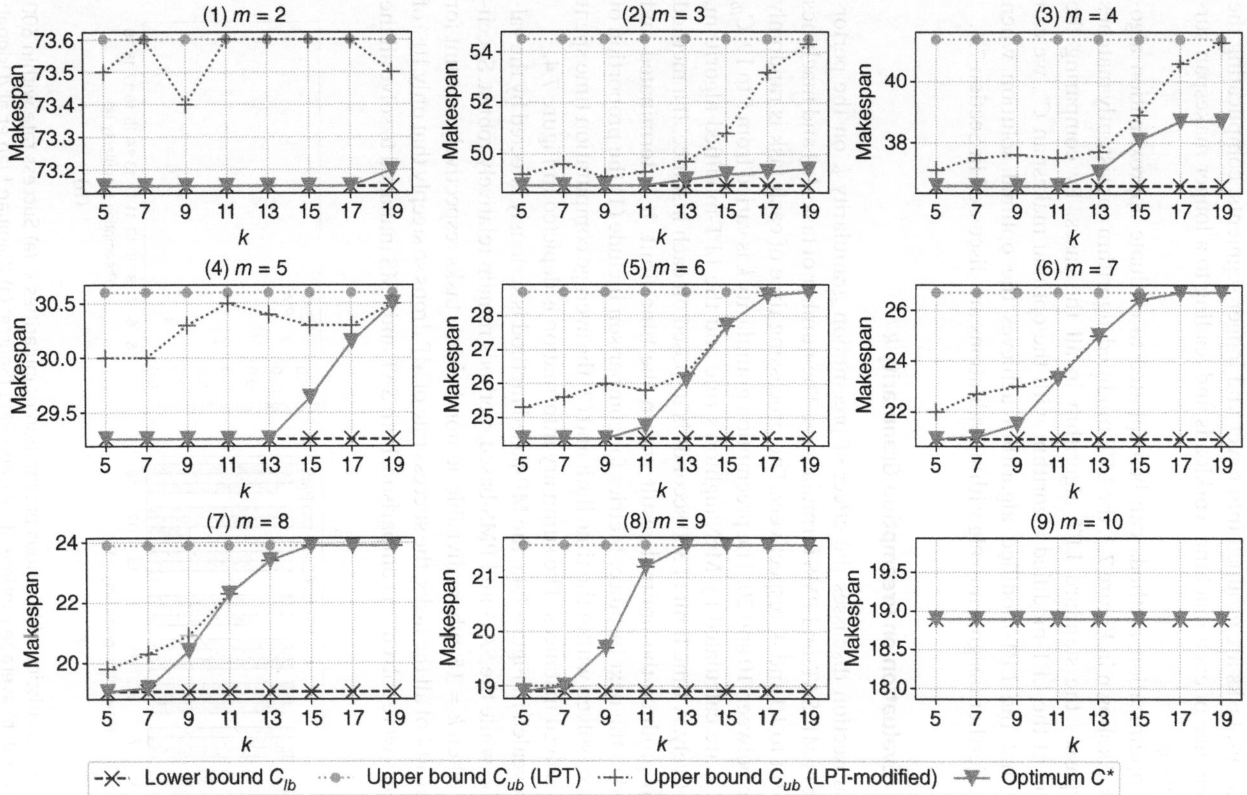

Figure 7.3 Optimal makespan under the setting of $n = 10$ and $p_j \sim DU(10, 20)$ (Source: Xiaojuan Liao et al. (2023)/with permission of Journal of Universal Computer Science).

task segmentation, ensuring uniform machine loading and facilitating $C^* = C_{lb}$. As k grows, tasks are grouped into fewer and longer segments, complicating the maintenance of even machine workloads and leading to a longer makespan surpassing C_{lb}.

The optimization methods can be exploited to evaluate approximation algorithms. As shown in Figure 7.3, the LPT-modified algorithm consistently matches or surpasses the standard LPT algorithm in all the cases. By comparing the solution of the LPT-modified algorithm with the optimal makespan C^*, we can assert that the LPT-modified algorithm achieves the optimal solution when $m = 9$. Further insights into algorithm evaluations are discussed in Section 7.6.

7.5.2 Evaluation on Preemption Granularity *k*

This subsection discusses the effect of preemption granularity k on the performance of MP, SAT, and PMS formulations. The number of tasks (n) and machines (m) is set to 10 and 4, respectively. The processing time of each task is randomly chosen between 10 and 20. The preemption granularity k is varied from 5 to 19. C_{lb} and C_{ub} are calculated by McNaughton's rule and the LPT-modified algorithm, respectively. A time limit of 300 seconds is imposed on each instance. If a method fails to produce the optimal result within the time limit, it is terminated and applied to the next instance. Metrics for comparison include (1) the proportion of instances solved within the time limit and (2) the average computation time spent on the solved instances. The summary information is depicted in Figure 7.4.

As revealed in Figure 7.4a, the MP-based method is seriously affected by the values of k, while the SAT- and PMS-based methods remain relatively robust. Specifically, when $k = 15$, where divisible ad nondivisible tasks respectively account for around 50% of all the tasks, the success rate of MP drops so steeply that only 10% of instances were solved. In comparison, both SAT and PMS managed to solve all the

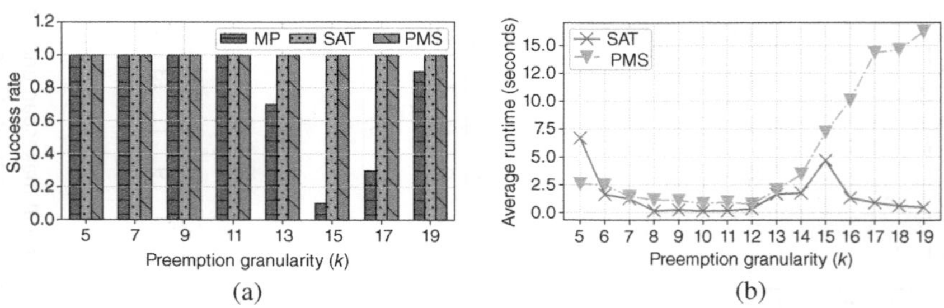

Figure 7.4 Statistics with various preemption granularities k. (a) Success rate within 300 seconds and (b) average runtime of SAT and PMS (Source: (a) Xiaojuan Liao et al./Orignal; (b) Xiaojuan Liao et al. (2023)/with permission of Journal of Universal Computer Science).

instances within the time limit. This indicates that satisfiability-based methods are more effective than MP when faced with varying preemption granularity k.

Figure 7.4b further compares the solving efficiency of SAT and PMS. Generally, when $k \leq 11$, the performance of both SAT and PMS improves with increasing k. However, when $k \geq 12$, the average runtime of SAT fluctuates, whereas PMS's runtime increases significantly as k rises. The following analysis examines the factors that impact the performance of SAT and PMS when $k \leq 11$ and $k > 11$.

When k is small enough, i.e., $k \leq 11$, tasks can be divided into smaller segments, ensuring a balanced machine load schedule. In this scenario, the SAT-based method only needs to call the solver once to verify that C_{lb} is adequate to satisfy all clauses. Similarly, PMS can also satisfy all clauses, both hard and soft. Therefore, the solving efficiency of SAT and PMS primarily depends on the time required to find a SAT solution, which is influenced by the search space and its constraints, represented by the number of variables and clauses.

When k exceeds 11, achieving a load-balanced schedule is no longer possible, and the optimal makespan C^* deviates from the lower bound C_{lb}. In SAT, to accurately determine the minimum makespan that satisfies the formula, the solver must be called repeatedly, indicating that not all clauses can be satisfied when the tested makespan is between C_{lb} and $C^* - 1$. To demonstrate that all clauses are satisfiable (i.e., reporting satisfiability), the solver needs only to find one feasible solution and stop searching once it is identified. In contrast, to declare that not all clauses are satisfiable (i.e., reporting unsatisfiability), the solver must explore the entire search space to ensure no variable assignment can satisfy the formula. Evidently, reporting unsatisfiability is more time consuming than finding a satisfiable solution within the same search space, and it has a greater impact on the solving efficiency of the SAT-based method.

Similarly, in PMS, when $k > 11$, not all soft clauses $z_t(C_{lb} \leq t \leq C_{ub})$ can be satisfied if $C^* > C_{lb}$. Solving instances with $C^* > C_{lb}$ requires identifying the minimum number of unsatisfied soft clauses, whereas solving those with $C^* = C_{lb}$ only requires finding a feasible variable assignment that satisfies all clauses. Based on the above analysis, solving instances with $C^* > C_{lb}$ takes significantly more time than those where $C^* = C_{lb}$, making it the dominant factor to affect the PMS's solving efficiency.

Tables 7.3 and 7.4 illustrate the key factors affecting the performance of SAT and PMS, documenting how these factors vary with increasing preemption granularity k. For more detailed comparisons, please refer to [5].

7.5.2.1 Evaluation on Number of Machines m

This section investigates the effect of an increased number of machines m. The number of tasks n is set to 10 and the preemption granularity k is fixed at 9. The processing time of each task is randomly chosen from 10 to 20. The value of

Table 7.3 The variation trend of factors influencing SAT performance as k increases.

Value of k	No. variables	No. clauses	No. UNSAT reports	Time per UNSAT report	Efficiency
$k \leq 11$	Decrease	Decrease	0	0	Increase[a)]
$k > 11$	Decrease	Decrease	Increase	Decrease	Fluctuate[b)]

a) Decreasing the numbers of variables and clauses enhance solving efficiency.
b) Reporting UNSAT is substantially more time consuming than reporting SAT, and the efficiency is influenced by both the number of UNSAT reports and the time taken per UNSAT report.

Table 7.4 The variation trend of factors influencing PMS performance as k increases.

Value of k	No. variables[a)]	No. clauses	No. instances with $C^* > C_{lb}$	Efficiency
$k \leq 11$	Fluctuate	Increase	0	Increase[b)]
$k > 11$	Increase	Increase	Increase	Decrease[c)]

a) The quantity of variables is directly influenced by C_{lb} and C_{ub}. When both C_{lb} and C_{ub} are unaffected by k, the variable count remains constant. The fluctuation in the number of variables is due to C_{ub}, which varies with k as determined by the LPT-modified algorithm.
b) Variable fluctuations indicate a restricted search scope, while increased clauses effectively guide the solver to prune the search space, thereby enhancing computational efficiency.
c) Solving instances where $C^* > C_{lb}$ consumes substantially more time compared to instances where $C^* = C_{lb}$, and is the dominant factor influencing the overall efficiency.

the critical parameter m is varied from 2 to 9. Comparison results are shown in Figure 7.5.

As shown in Figure 7.5a, the success rate of PMS remains consistently at 1 across all machine numbers, while the success rates of MP and SAT fluctuate when $7 \leq m \leq 9$. In this range, SAT managed to solve more instances than MP. This demonstrates that the satisfiability-based methods are more effective than MP across varying numbers of machines.

Figure 7.5b illustrates the average runtime of SAT and PMS for different values of m. A number with an arrow in the figure indicates the proportion of instances solved within 300 seconds, which is omitted if the solver successfully addressed all tested instances. Clearly, the efficiency of SAT fluctuates significantly, while that of PMS remains relatively stable with the increasing m.

That $m = 6$ is a watershed. The main factors impacting performance of SAT and PMS are different when $m \leq 6$ and $m > 6$, depending on whether C^* is equal to C_{lb}. Tables 7.5 and 7.6 show the main factors influencing the performance of SAT

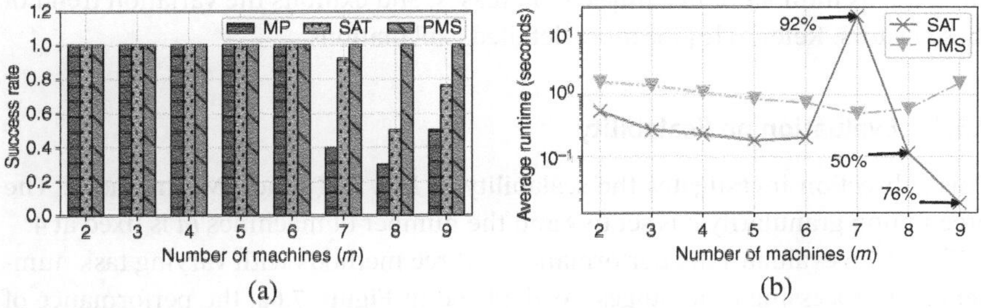

Figure 7.5 Statistics with various numbers of machines m. (a) Success rate within 300 seconds and (b) average runtime of SAT and PMS (Source: (a) Xiaojuan Liao et al./Orignal; (b) Xiaojuan Liao et al. (2023)/with permission of Journal of Universal Computer Science).

Table 7.5 The variation trend of factors influencing SAT performance as m increases.

Value of m	No. variables	No. clauses	No. UNSAT reports	Time per UNSAT report	Efficiency
$m \leq 6$	Decrease	Increase	0	0	Increase[a)]
$m > 6$	Increase	Increase	Fluctuate	Increase	Fluctuate[b)]

a) Reducing the number of variables narrows the search space, while increasing the number of clauses adds more constraints, helping the solver find a feasible variable assignment more quickly.
b) Reporting UNSAT is substantially more time consuming than reporting SAT, and the efficiency is influenced by both the number of UNSAT reports and the time taken per UNSAT report.

Table 7.6 The variation trend of factors influencing PMS performance as m increases.

Value of m	No. variables[a)]	No. clauses	No. instances with $C^* > C_{lb}$	Efficiency
$m \leq 6$	Decrease	Decrease	0	Increase[a)]
$m > 6$	Increase	Increase	Increase	Decrease[b), c)]

a) Decreasing the numbers of variables and clauses enhance solving efficiency.
b) Solving instances where $C^* > C_{lb}$ consumes substantially more time compared to instances where $C^* = C_{lb}$, and is the dominant factor influencing the overall efficiency.
c) Increasing numbers of variables and clauses decrease solving efficiency.

and PMS as number of machines m increases, and exhibits the variation trend of these factors. Refer to [5] for more detailed explanations.

7.5.3 Evaluation on Scalability

This subsection investigates the scalability of MP, SAT, and PMS methods. The preemption granularity k is set to 9 and the number of machines m is fixed at 4.

Figure 7.6 evaluates the performance of three methods with varying task numbers and processing time ranges. As depicted in Figure 7.6a, the performance of all techniques degrades as the number of tasks (n) increases from 10 to 30. Among these methods, MP demonstrates superior performance, solving all instances with an average runtime that moderately increases from around 1 second to 40 seconds. In contrast, the solving efficiency of satisfiability-based methods deteriorates significantly, with no instances being solved when n reaches 25. This indicates that SAT and PMS techniques are less suitable than MP for handling a large number of tasks.

Figure 7.6b compares performance across varying processing time ranges $[t, 2t]$, with t ranging from 10 to 50. As t increases, the success rate of PMS drops rapidly, failing to solve any instances when t reaches 40. In contrast, MP and SAT maintain a constant success rate of 1 regardless of changes in t. Specifically, the average runtime of SAT increases gradually from around 0.2 to 2 seconds as t rises from 10 to 50, while MP solves all instances in about 1 second.

The performance of MP and SAT is further compared across a broader range of processing times. According to [6], problems can be optimally solved without preemption if tasks have widely varying lengths. In this extended experiment, instead of varying the processing time from t to $2t$, a narrower range of $[t, t + 30]$ is used, with t changing from 60 to 120 in increments of 10. Figure 7.7 illustrates the comparison results. As t increases from 60 to 100, the success rate of MP drops from

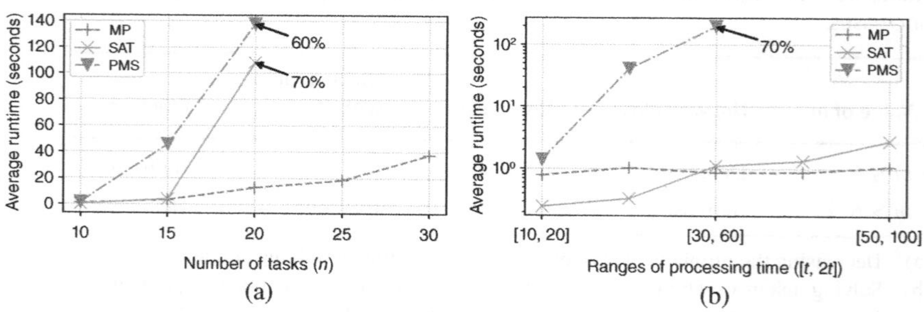

Figure 7.6 Performance evaluation on scalability. (a) Average runtime with various number of tasks where $p_j \sim DU(10, 20)$ and (b) average runtime with various ranges of processing time where $n = 10$ (Source: Xiaojuan Liao and Miyuki Koshimura/original).

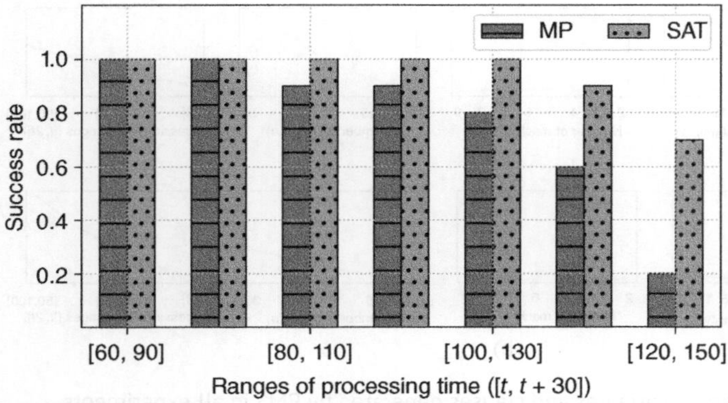

Figure 7.7 Performance comparison with MP and SAT with larger processing time (Source: Xiaojuan Liao and Miyuki Koshimura/original).

1 to 0.8, while SAT solves all instances within the 300-second time limit. When t reaches 120, PMS solves only 20% of instances,

Finally, let us examine the difference in performance between the SAT and PMS methods in the scalability experiments. In this subsection, the optimal makespan C^* of each instance tested is always equal to C_{lb}. Therefore, both SAT- and PMS-based methods only require their solvers to find a feasible variable assignment that satisfies all clauses. The computation time is primarily influenced by the search space and associated constraints, reflected in the number of Boolean variables and clauses.

Figures 7.8 and 7.9 display the numbers of variables and clauses generated by SAT and PMS. To better understand the results, we compare the problem scales

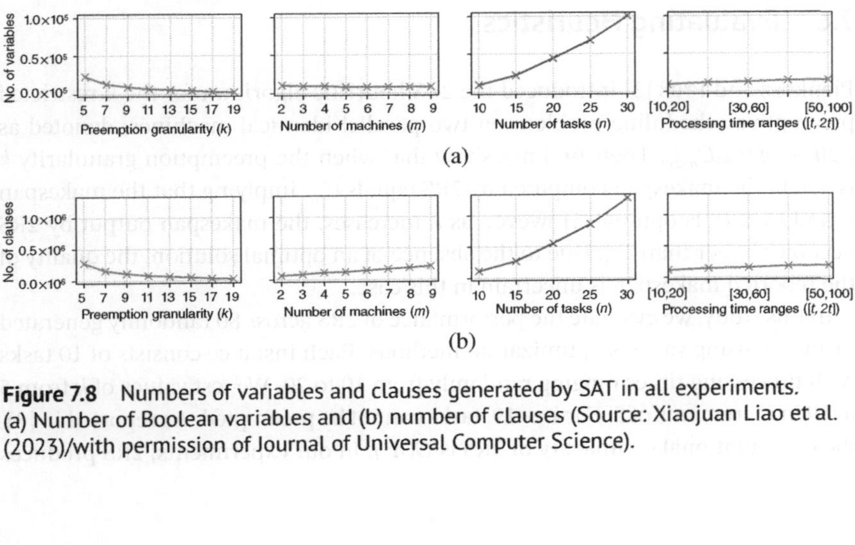

Figure 7.8 Numbers of variables and clauses generated by SAT in all experiments. (a) Number of Boolean variables and (b) number of clauses (Source: Xiaojuan Liao et al. (2023)/with permission of Journal of Universal Computer Science).

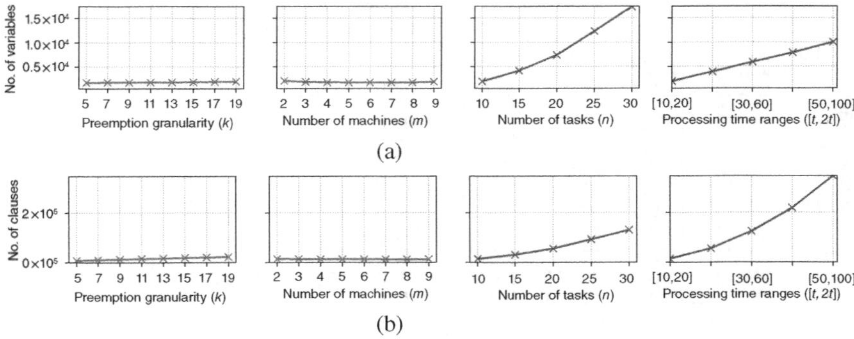

Figure 7.9 Numbers of variables and clauses generated by PMS in all experiments. (a) Number of Boolean variables and (b) number of clauses (Source: Xiaojuan Liao et al. (2023)/with permission of Journal of Universal Computer Science).

across all experiments. As shown in Figure 7.8, the number of variables and clauses increases dramatically with the number of tasks but remains relatively stable with the expansion of processing time ranges. This explains why SAT performance degrades significantly as the number of tasks increases but remains efficient with expanding processing time ranges. In contrast, Figure 7.9 shows that the number of variables and clauses generated by PMS increases noticeably with either the number of tasks or the processing time. Consequently, in the scalability experiments, PMS performance declines sharply with increasing task numbers and processing time ranges.

7.6 Evaluating Heuristics

Pieńkosz and Prus [3] introduced the 2RS heuristic algorithm for the k-restricted preemptive scheduling problem on two parallel identical machines, denoted as $P2|k-pmtn|C_{max}$. Their findings show that when the preemption granularity k is small, the makespan computed by 2RS equals C_{lb}, implying that the makespan found by 2RS is optimal. However, as k increases, the makespan output by 2RS becomes larger than C_{lb}. Due to the absence of an optimal solution, the quality of the reported makespan is uncertain in this case.

In this study, we evaluate the performance of 2RS across 80 randomly generated instances using various optimization methods. Each instance consists of 10 tasks with processing times ranging randomly from 10 to 20. We test values of k from 5 to 19 in increments of 2. For $k > 19$, tasks cannot be preempted. As reported in [3], the computational complexity of 2RS is $O(n^2)$. In our experiments, 2RS produces

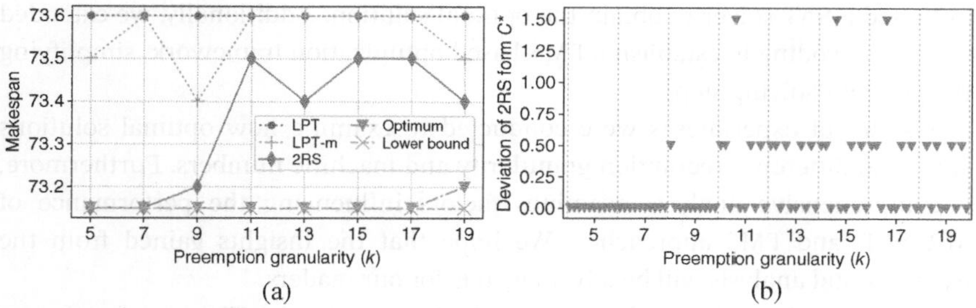

Figure 7.10 Evaluation of 2RS. (a) Average values of makespan for various values of k and (b) deviation of 2RS's solution from the optimal makespan (Source: Xiaojuan Liao et al. (2023)/with permission of Journal of Universal Computer Science).

approximate solutions in around 1.5 milliseconds, which is substantially faster than optimal methods.

Figure 7.10a presents the average makespan values obtained by heuristics for different k values. Each data point represents the average makespan of 10 instances. For better comparison, we include the lower bound computed by McNaughton's rule, and results from two classical heuristics, LPT and LPT-m, as shown in Figure 7.10a.

From Figure 7.10a, it is evident that 2RS consistently outperforms LPT and LPT-modified algorithms across different k values, but its performance degrades with larger k. Specifically, for small k ($k \leq 7$), 2RS achieves makespans equal to C^*, whereas for larger k ($k \geq 9$), the makespan exceeds C^*.

Figure 7.10b illustrates the deviation of 2RS solutions from C^* on each instance. When $k \leq 7$, 2RS finds C^* for all instances. However, for $k = 9$, 2RS fails to find C^* in one out of 10 instances, and for $k \geq 11$, it identifies C^* for about half of the instances. Despite not always achieving optimality, 2RS consistently produces high-quality approximation solutions, with the maximum deviation from C^* being just 1.5, demonstrating its effectiveness in practical applications.

7.7 Chapter Summary

This chapter explores the $P|k-pmtn|C_{max}$ problem, focusing on minimizing the makespan in parallel-machine k-restricted preemptive scheduling. Known for its NP-hardness, many studies have aimed to develop approximation algorithms to find suboptimal solutions.

To achieve optimal solutions, we first introduced a MP model tailored to address the $P|k-pmtn|C_{max}$ problem. Building on advancements in SAT-solving techniques, we encoded the MP model into propositional logical formulas and exploited a SAT solver to obtain the optimal solution. Additionally, we extended the SAT encoding to establish a PMS-based optimization framework, simplifying the problem-solving steps.

A series of experiments were conducted to examine how optimal solutions vary with different preemption granularity and machine numbers. Furthermore, a comprehensive analysis examines factors influencing the performance of MP, SAT, and PMS approaches. We hope that the insights gained from the experimental analysis will be advantageous for our readers.

Lastly, an existing heuristic algorithm 2RS was evaluated. The use of optimization methods enables a more precise assessment of the quality of solutions produced by heuristic algorithms.

References

1 R. L. Graham, E. L. Lawler, J. K. Lenstra, and A. H. G. Rinnooy Kan. Optimization and approximation in deterministic sequencing and scheduling: a survey. *Annals of Discrete Mathematics*, 5(1): 287–326, 1979.

2 K. Ecker and R. Hirschberg (1993). Task scheduling with restricted preemptions. In Arndt Bode, Mike Reeve, and Gottfried Wolf, editors, *PARLE '93 Parallel Architectures and Languages Europe*, pages 464–475, Berlin, Heidelberg, 1993. Springer.

3 K. Pieńkosz and A. Prus. Task scheduling with restricted preemptions on two parallel processors. In *2015 20th International Conference on Methods and Models in Automation and Robotics (MMAR)*, pages 58–61, Miedzyzdroje, Poland, Aug. 2015. IEEE.

4 S. Knust, N. Shakhlevich, Stefan Waldherr, and C. Weiß. Shop scheduling problems with pliable jobs. *Journal of Scheduling*, 22(6): 635–661, 2019.

5 Xiaojuan Liao, Hui Zhang, Miyuki Koshimura, Rong Huang, and Fagen Li. Solving restricted preemptive scheduling on parallel machines with SAT and PMS. *Journal of Universal Computer Science*, 29(8): 911–937, 2023.

6 T. Barański. Task scheduling with restricted preemptions. In *2011 Federated Conference on Computer Science and Information Systems (FedCSIS)*, pages 231–238, Szczecin, Poland, Nov. 2011. IEEE.

7 R. McNaughton. Scheduling with deadlines and loss functions. *Management Science*, 6: 1–11, 1959.

8 Olivier Roussel and Vasco M. Manquinho. Pseudo-Boolean and cardinality constraints. In Armin Biere, Marijn Heule, Hans van Maaren, and Toby Walsh, editors, *Handbook of Satisfiability* - Second Edition, volume 336 of *Frontiers in Artificial Intelligence and Applications*, pages 1087–1129. IOS Press, 2021.

9 Alexey Ignatiev, Antonio Morgado, and Joao Marques-Silva. PySAT: A Python toolkit for prototyping with SAT oracles. In *International Conference on Theory and Applications of Satisfiability Testing*, pages 428–437, Cham, 2018. Springer International Publishing

10 LLC Gurobi Optimization. Gurobi optimizer reference manual, 2021.

11 Gilles Audemard, Jean-Marie Lagniez, and Laurent Simon. Improving glucose for incremental SAT solving with assumptions: Application to MUS extraction. In *Theory and Applications of Satisfiability Testing – SAT 2013*, pages 309–317, Berlin, Heidelberg, 2013. Springer-Verlag.

12 N. Eén and N. Sörensson. An extensible SAT-solver. In *International Conference on Theory and Applications of Satisfiability Testing*, pages 502–518, Italy, May 2003. Springer.

13 A. Morgado, F. Heras, M. H. Liffiton, J. Planes, and Marques-Silva J. Iterative and core-guided MaxSAT solving: A survey and assessment. *Constraints*, 18(4): 478–534, 2013.

14 Mark H. Liffiton and Jordyn C. Maglalang. A cardinality solver: More expressive constraints for free. In *Proceedings of the 15th International Conference on Theory and Applications of Satisfiability Testing*, SAT'12, page 485–486, Berlin, Heidelberg, 2012. Springer-Verlag.

8
Rule Relation-Based Weighted Partial MaxSAT (RWPM) Encoding

After reading this chapter, you should be able to:

- Define the coalition structure generation (CSG) problem in characteristic function games and partition function games.
- Describe the CSG problem in compact representation schemes marginal contribution networks (MC-nets) and embedded MC-nets.
- Apply weighted partial MaxSAT (WPM) to solve the CSG problem based on rule relations among MC-nets and embedded MC-nets.

8.1 Problem Statement

Coalition formation is a crucial capability in multiagent systems, involving agents collaborating and coordinating joint actions to accomplish a set of tasks collectively. Sample applications of coalition formation include distributed vehicle routing [1], sensor network [2], and e-commerce [3]. In general, coalition formation is composed of three main activities [4]:

- **Coalition structure generation**: Finding a coalition structure, i.e., an exhaustive set of mutually disjoint coalition, so that the performance of the entire system is optimized.
- **Teamwork**: Optimizing the performance of each individual coalition.
- **Payoff distribution**: Dividing the gains from cooperation among agents so as to meet certain positive or normative criteria.

Coalition structure generation (CSG) is a primary challenge in coalition formation within multiagent systems [5]. The CSG problem involves partitioning a set of agents into exhaustive and disjoint coalitions, where each coalition is assigned a real-valued payoff. This partition is called a coalition structure. The objective of

Applied Satisfiability: Cryptography, Scheduling, and Coalitional Games, First Edition. Xiaojuan Liao and Miyuki Koshimura.

solving the CSG problem is to identify a coalition structure that maximizes the total value of all the coalitions.

Most existing works assume that a coalition's value is independent of other coalitions in the coalition structure. These settings are referred to as characteristic function games (CFGs), where a coalition's value is determined by a *characteristic function*. Many real-world multiagent problems are instances of CFGs [1, 5]. For example, universities and research institutions often collaborate on large projects. The value of a coalition is determined by the combined expertise and resources of the institutions involved. The goal is to form coalitions that maximize research output and funding. In an edge computing environment, edge nodes often encounter the computing scenario of multiple tasks arriving simultaneously. Given their limited resources, a single node cannot handle all tasks concurrently. Therefore, cooperation between nodes is essential, with the objective being to form coalitions of nodes to effectively solve the multitask concurrency problem.

A more realistic form of coalition values, called *partition function*, assumes that a coalition's value is influenced by the formation of other coalitions. This type of coalitional game is known as partition function games (PFGs). The impact of other coalitions on a coalition's value is referred to as externality from coalition formation. For instance, companies in the telecommunications industry might form coalitions to share infrastructure. The benefit of a coalition depends on how the rest of the market is structured, such as the presence of other competing coalitions and market coverage. More examples of games with externalities, such as collusion in oligopolies, congestion games, various forms of international policy coordination between countries are covered in [6, 7]. Concrete examples of the CSG problem in CFGs and PFGs are shown in Examples 8.1 and 8.2, respectively.

Example 8.1 Consider coalition formation among three autonomous agent-robots a_1, a_2, and a_3. The payoffs of possible coalitions are listed as follows:

- If a_1 works alone, the payoff of 1 is achieved.
- If a_2 works alone, the payoff of 0 is achieved.
- If a_3 works alone, the payoff of 0 is achieved.
- If a_1 and a_2 work together, the payoff of 1 is achieved.
- If a_1 and a_3 work together, the payoff of 1 is achieved.
- If a_2 and a_3 work together, the payoff of 1 is achieved.
- If a_1, a_2 and a_3 work together, the payoff of 2 is achieved.

It indicates that a_1 is able to secure the payoff of 1 if acting alone, while a_2 and a_3 produce nothing if performing separately. Besides, no pair of cooperating agents achieves any value added except a_2 and a_3 who are able to lead to the payoff of 1 if they work together. There are two optimal partitions of these three agents. The first one is the grand coalition, i.e., a_1, a_2, and a_3 perform cooperatively. The other one is to assign a_2 and a_3 to the same group while let a_1 work alone. The total values of these two partitions are both 2.

Example 8.2 Consider coalition formation among the three autonomous agent-robots a_1, a_2, and a_3 again. The payoffs of coalitions are almost the same as that in Example 8.1, except that the following condition is incorporated.

- The performance of a_1 improves by an additional 0.5 when facing the cooperation of a_2 and a_3.

In this case, the value of coalition $\{a_1\}$ is different in $\{\{a_1\}, \{a_2, a_3\}\}$ than in $\{\{a_1\}, \{a_2\}, \{a_3\}\}$. When considering the payoff of $\{a_1\}$, the formation of other coalitions should be taken into account, i.e., whether a_2 and a_3 are assigned to the same coalition.

These two examples provide an overview of two types of the CSG problem. In Example 8.1, the value of each coalition is independent of how other coalitions are formed, while in Example 8.2, the value of coalition $\{a_1\}$ is affected by the formation of other coalitions, which embodies *externality*.

Given a set of agents A, a coalition, denoted by C, is a nonempty subset of A, i.e., $C \in 2^A \backslash \emptyset$. A *coalition structure*, CS, is an exhaustive set of mutually disjoint coalitions over A, i.e., CS is subject to the constraints: $\forall i, j\,(i \neq j), C_i \cap C_j = \emptyset, \bigcup_{C_i \in CS} C_i = A$. We denote by $\Pi(A)$ the set of all coalition structures over A.

8.1.1 Characteristic Function Game

In a setting without externalities, the value of a coalition C is given by a characteristic function $v : 2^A \to \mathbb{R}$, assigning a real-valued payoff to each coalition $C \subseteq A$. The value of a coalition structure CS is called *social welfare*, denoted as $V(CS)$, given by $V(CS) = \sum_{C_i \in CS} v(C_i)$. The objective of solving the CSG problem is to find an optimal coalition structure that makes the social welfare maximized, i.e., given A, find CS^* such that $\forall CS \in \Pi(A), V(CS^*) \geq V(CS)$.

Example 8.3 Reconsider the game in Example 8.1. For $A = \{a_1, a_2, a_3\}$, the characteristic function is denoted as:

$$v(\{a_1\}) = 1, \quad v(\{a_2\}) = 0, \quad v(\{a_3\}) = 0,$$
$$v(\{a_1, a_2\}) = 1, \quad v(\{a_1, a_3\}) = 1, \quad v(\{a_2, a_3\}) = 1,$$
$$v(\{a_1, a_2, a_3\}) = 2$$

There are five possible coalition structures. Figure 8.1 illustrates the enumeration of agents within these coalition structures. The corresponding social welfare

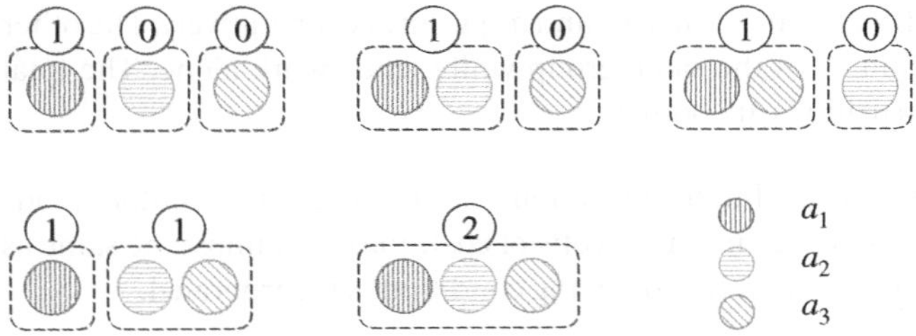

Figure 8.1 Enumeration of agents within all possible coalition structures in Example 8.3. Each number in a circle represents the value of the corresponding coalition (Source: Xiaojuan Liao and Miyuki Koshimura/original).

values are listed below.

$$V\left(\{\{a_1\},\{a_2\},\{a_3\}\}\right)=1, \qquad V\left(\{\{a_1,a_2\},\{a_3\}\}\right)=1,$$
$$V\left(\{\{a_1,a_3\},\{a_2\}\}\right)=1, \qquad V\left(\{\{a_1\},\{a_2,a_3\}\}\right)=2,$$
$$V\left(\{\{a_1,a_2,a_3\}\}\right)=2$$

It is clear that there are two optimal coalition structures: $\{\{a_1\},\{a_2,a_3\}\}$ and $\{\{a_1,a_2a_3\}\}$, with the social welfare of 2.

The difficulty of solving the CSG problem lies in the exponentially increasing number of possible solutions (i.e., coalition structures), with $O(n^n)$ for n agents. Moreover, naive representation of characteristic functions enumerates the payoffs to each possible set of agents, requiring space exponential in the number of agents. To be more specific, the representation size the CSG problem is $O(2^n)$ for CFGs, where n is the number of agents. This is prohibitive for large n. Ieong and Shoham [8] developed a concise representation called marginal contribution network (MC-net), which largely reduces the space necessary for representation.

Definition 8.1 ***(MC-nets)*** An MC-net consists of a set of *rules* R. Each rule $r_i \in R$ is expressed in a syntactic form $I_i \rightarrow w_i$, where $w_i \in \mathbb{R}$ and I_i is the condition of rule r_i, denoted by a conjunction of literals over A, i.e., $\{a_1 \wedge a_2 \wedge \cdots \wedge a_l \wedge \neg a_{l+1} \wedge \cdots \wedge \neg a_m\}$. For each rule r_i, P_i is called *positive literals* where $P_i = \{a_j\}_{j=1}^{l}$, and N_i is called *negative literals* where $N_i = \{a_j\}_{j=l+1}^{m}$. A rule r_i is said to *apply to* a coalition C if $P_i \subseteq C$ and $N_i \cap C = \emptyset$, i.e., all agents in P_i are in C, and none of agents in N_i are in C. For a coalition C, $v(C) = \sum_{r_i \in R'} w_i$, where R' is the set of rules that apply to C. Thus, the value of a coalition structure CS is given as $\sum_{C \in CS} v(C)$. Without loss of generality, it is assumed that each rule has at least one positive literal.

Thus, in MC-nets, the problem of finding CS^* is equivalent to finding a set of rules that apply to some coalition $C \in CS^*$, such that the sum of values in the set of rules is maximized.

In practice, the procedure for checking whether a rule $r : (I_r \to w_r)$ applies to a coalition C can proceed in the following way. First assign *true* to all the agents present in C and *false* to others. Based on such assignment, examine the truth value of I_r. If I_r evaluates to true, the rule r applies to C and the value of w_r is obtained. Otherwise, r does not apply to C and the corresponding gain is zero.

Example 8.4 The game in Example 8.3 can be represented with just two rules:

- $r_1 : a_1 \to 1$,
- $r_2 : a_2 \wedge a_3 \to 1$.

r_1 applies to coalition $\{a_1\}, \{a_1, a_2\}, \{a_1, a_3\}$, and $\{a_1, a_2, a_3\}$. r_2 applies to $\{a_2, a_3\}$ and $\{a_1, a_2, a_3\}$. Both rules apply to $\{\{a_1\}, \{a_2, a_3\}\}$ and $\{\{a_1, a_2, a_3\}\}$, thus there are two optimal coalition structures: $\{\{a_1\}, \{a_2, a_3\}\}$ and $\{\{a_1, a_2, a_3\}\}$. The maximized social welfare is $V(CS^*) = 2$.

8.1.2 Partition Function Game

In a setting with externalities, an *embedded coalition* is a pair (C, CS). Let M denote the set of all embedded coalitions, i.e., $M := \{(C, CS) : CS \in \Pi(A), C \in CS\}$. The value of a coalition depends on the formation of other coexisting coalitions in the coalition structure, and is specified by a partition function, which is a mapping $w : M \to \mathbb{R}$. A game in a partition function form is a tuple (A, w). The problem of solving CSG seeks a coalition structure CS, such that $V(CS^*) = \sum_{C \in CS^*} w(C, CS^*)$ is optimized.

Example 8.5 Reconsider the game in Example 8.2. For $A = \{a_1, a_2, a_3\}$, the partition function is denoted as:

$$\begin{aligned}
&w(\{a_1\}, \{\{a_1\}, \{a_2\}, \{a_3\}\}) = 1, && w(\{a_1\}, \{\{a_1\}, \{a_2, a_3\}\}) = 1.5,\\
&w(\{a_2\}, \{\{a_1\}, \{a_2\}, \{a_3\}\}) = 0, && w(\{a_2\}, \{\{a_2\}, \{a_1, a_3\}\}) = 0,\\
&w(\{a_3\}, \{\{a_1\}, \{a_2\}, \{a_3\}\}) = 0, && w(\{a_3\}, \{\{a_3\}, \{a_1, a_2\}\}) = 0,\\
&w(\{a_1, a_2\}, \{\{a_1, a_2\}, \{a_3\}\}) = 1, && w(\{a_1, a_3\}, \{\{a_2\}, \{a_1, a_3\}\}) = 1,\\
&w(\{a_2, a_3\}, \{\{a_1\}, \{a_2, a_3\}\}) = 1, && w(\{a_1, a_2, a_3\}, \{\{a_1, a_2, a_3\}\}) = 2
\end{aligned}$$

There are five possible coalition structures. Figure 8.2 illustrates the enumeration of agents within these coalition structures. The corresponding social welfare values are listed below.

$$\begin{aligned}
&V(\{\{a_1\}, \{a_2\}, \{a_3\}\}) = 1, && V(\{\{a_1, a_2\}, \{a_3\}\}) = 1,\\
&V(\{\{a_1, a_3\}, \{a_2\}\}) = 1, && V(\{\{a_1\}, \{a_2, a_3\}\}) = 2.5,\\
&V(\{\{a_1, a_2, a_3\}\}) = 2
\end{aligned}$$

By enumerating all possible coalition structures and comparing their social welfare, one can conclude that the optimal coalition structure is $\{\{a_1\}, \{a_2, a_3\}\}$, with the social welfare of 2.5.

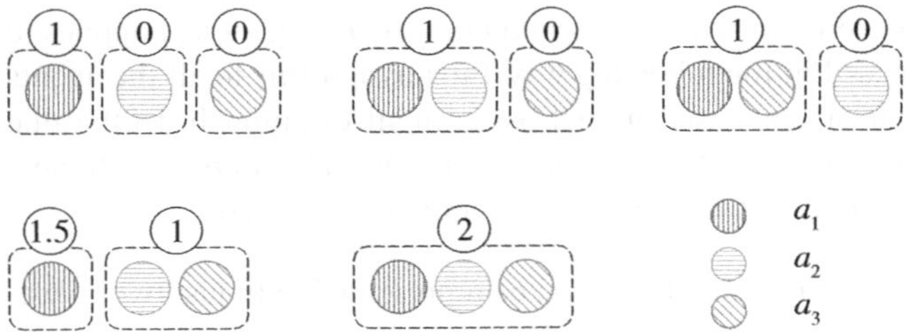

Figure 8.2 Enumeration of agents within all possible coalition structures in Example 8.5. Each number in a circle represents the value of the corresponding coalition (Source: Xiaojuan Liao and Miyuki Koshimura/original).

Coalitional games with externalities are more complex since the value depends not only on individual coalitions, but also on embedded coalitions. The representation size increases to $O(n^n)$ for PFGs [9]. Therefore, solving the CSG problem for PFGs is more challenging than the CFG case, which is already NP-complete [5]. Michalak et al. [10] develop a concise representation scheme for PFGs by extending MC-nets to a partition function, called embedded MC-nets.

Definition 8.2 ***(Embedded MC-nets)*** An embedded MC-net is given by a set of embedded rules ER. Each rule $er \in ER$ is expressed in a syntactic form $I_0|I_1, \dots, I_k \rightarrow w_{er}$, where each $I_i : i \in \{0, \dots, k\}$ is conjunction of literals over A. A rule er is said to *apply to* an embedded coalition (C, CS) if

1) I_0 applies to C, and
2) each $I_i : i \in \{1, \dots, k\}$ applies to at least one coalition in $CS \setminus \{C\}$.

For an embedded coalition (C, CS), $w(C, CS) = \sum_{er \in ER'} w_{er}$, where ER' is the set of embedded rules that apply to (C, CS).

Example 8.6 The game with externalities in Example 8.5 can be described with the following set of rules:

- $r_1 : a_1 \rightarrow 1$,
- $r_2 : a_2 \wedge a_3 \rightarrow 1$,
- $r_3 : a_1 | a_2 \wedge a_3 \rightarrow 0.5$.

r_1 applies to coalition $\{a_1\}$, $\{a_1, a_2\}$, $\{a_1, a_3\}$, and $\{a_1, a_2, a_3\}$. r_2 applies to $\{a_2, a_3\}$ and $\{a_1, a_2, a_3\}$. r_3 applies to an embedded coalition $(\{a_1\}, \{\{a_1\}, \{a_2, a_3\}\})$. All rules apply to $\{\{a_1\}, \{a_2, a_3\}\}$, thus the optimal coalition structure is $\{\{a_1\}, \{a_2, a_3\}\}$ and the optimized social welfare is $V(CS^*) = V(\{\{a_1\}, \{a_2, a_3\}\}) = 2.5$.

It worth noting that in this example, $\{a_1, a_2, a_3\}$ is no longer the optimal solution because in this case, r_3 does not apply to the grand coalition and the corresponding payoff of 0.5 cannot be obtained.

In Example 8.6, r_1 and r_3 highlight the difference between games with and without externalities. In r_1, agent a_1 contributes to any embedded coalition it belongs with value 1. Additionally, r_3 says that if there coexists another embedded coalition where a_2 and a_3 cooperate, the value of a_1 is increased by 0.5. This happens in $\{\{a_1\}, \{a_2, a_3\}\}$. In this way, embedded MC-nets allow to capture externalities in coalitional games [10].

8.2 Representative Algorithms

This section begins with an overview of algorithm classifications for solving the CSG problem and then details a representative algorithm known as direct encoding [11, 12], which addresses the CSG problem in both CFGs and PFGs. This direct encoding approach serves as the foundation for the WPM encoding based on rule relations, which will be introduced in Section 8.3.

8.2.1 An Overview

There have been a number of approaches to solving the coalition structure generation (CSG) problem, classified into dynamic programming, anytime optimal algorithms, and heuristic approximate algorithms. Their perspective properties are summarized as follows:

- Dynamic programming algorithms, first introduced by Yeh [13], break down the problem into smaller subproblems, solving them recursively, and combining their results to form a final solution. Despite providing optimal solutions with relatively low worst complexity, dynamic programming algorithms require evaluating all possible agent combinations, leading to high memory usage and redundant calculations.
- Anytime optimal algorithms, like integer partition [14], can return intermediate solutions if interrupted, improving solution quality as computation time increases. These algorithms also offer worst-case guarantees on solution quality.
- For larger problems, heuristics such as tabu search, simulated annealing [15], genetic algorithms [16], ant optimization [17], and particle swarm algorithms [18] are used. These algorithms do not guarantee optimal solutions or worst-case performance but are computationally tractable for large-scale problems.

Another research direction focuses on solving CSG with concise representations [11, 12, 19, 20]. In this way, the representation size can be remarkably reduced.

In this context, Ohta et al. [11] developed efficient constraint optimization algorithms for solving CSG in CFGs, demonstrating that the use of a concise representation MC-net significantly reduces computation time. For PFGs, where space efficiency is critical due to their larger size in naive representations, the embedded MC-net [10] was introduced as a compact representation. Building upon the formalization by Ohta et al. [11], Ueda et al. [12] developed a direct encoding method capable of handling the CSG problem with externalities. Subsequently, various weighted partial MaxSAT (WPM) encoding techniques were proposed [21–25]. These techniques translate the CSG problem, represented in compact form, into Boolean propositional logic, significantly expanding the scope of solvable problem instances.

8.2.2 Revisiting Important Works

In (embedded) MC-nets, each (embedded) rule is associated with a real-valued payoff. If the value is positive, the rule is called a positive value (embedded) rule. If the value is negative, the rule is called a negative value (embedded) rule. This section reviews important existing works on handling positive and negative value rules, which form the basis for the rule relation-based WPM encoding introduced in this chapter. The discussion on embedded rules will be addressed in Sections 8.3.2 and 8.3.4.

Definition 8.3 ***(Feasible rule set [11])*** A set of rules R is *feasible* if there exists a coalition structure CS where each rule R applies to some coalition C where $C \in CS$.

Thereby, the problem of finding CS^* is equivalent to finding a feasible rule set, such that the sum of values of the rule set is maximized.

To handle positive value rules in MC-nets, Ohta et al. [11] represent an MC-net in a rule relations-based graph. In this graph, each vertex is a rule, and between any two vertices, there exists an edge whose type is one of the four cases described in Definition 8.4.

Definition 8.4 ***(Relations between rules [11])*** The possible relations between rules r_i and r_j are classified into the following four nonoverlapping and exhaustive cases:

- **Compatible on the same coalition**: $P_i \cap P_j \neq \emptyset$ and $P_i \cap N_j = P_j \cap N_i = \emptyset$.
- **Compatible on different coalitions**: $P_i \cap P_j = \emptyset$ and ($P_i \cap N_j \neq \emptyset$ or $P_j \cap N_i \neq \emptyset$).

- **Incompatible**: $P_i \cap P_j \neq \emptyset$ and ($P_i \cap N_j \neq \emptyset$ or $P_j \cap N_i \neq \emptyset$).
- **Independent**: $P_i \cap P_j = \emptyset$ and $P_i \cap N_j = P_j \cap N_i = \emptyset$.

Example 8.7 *([11])*
Let there be five agents a_1, a_2, a_3, a_4, a_5 and four rules:

$$r_1 : a_2 \wedge a_5 \rightarrow 3, \qquad r_2 : a_1 \wedge a_2 \wedge a_3 \wedge \neg a_4 \rightarrow 2,$$
$$r_3 : a_1 \wedge a_4 \rightarrow 1, \qquad r_4 : a_3 \wedge \neg a_5 \rightarrow 1$$

The relations among these rules are listed as follows:

- r_1 and r_2 are compatible on the same coalition.
- r_1 and r_4 are compatible on different coalitions.
- r_2 and r_3 are incompatible.
- r_2 and r_4 are compatible on the same coalition.
- Other unmentioned rule relations are independent.

The graph representation of this example is depicted in Figure 8.3 [11], where the condition of each rule is denoted in the form of (P_i, N_i). For instance, $(\{a_1, a_2, a_3\}, \{a_4\})$ represents the condition of r_1, i.e., $a_1 \wedge a_2 \wedge a_3 \wedge \neg a_4$.
Theorem 8.1 [11] is proved to be sufficient to find out a feasible rule set among positive value rules in MC-nets. Refer readers to [11] for the proof of the theorem.

Theorem 8.1 *([11])* A set of rules R' is feasible if and only if R' satisfies the following conditions:

1) R' includes no pair of rules/vertices connected by an "incompatible" edge, and
2) if two rules in R' have the relation of "compatible on different coalitions," then they are not reachable via "compatible on the same coalition" edges in R'.

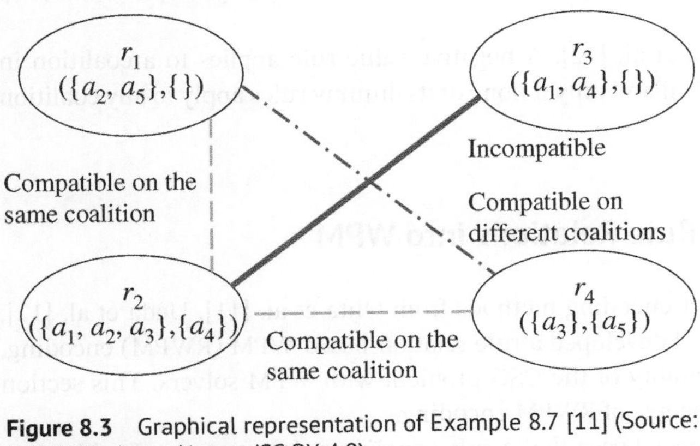

Figure 8.3 Graphical representation of Example 8.7 [11] (Source: Naoki Ohta et al. (2009)/Springer Nature/CC BY 4.0).

Example 8.8 Solving the problem introduced in Example 8.7 involves finding a feasible rule set that maximizes the sum of the rule values. Since all rules have positive values, selecting as many rules as possible is desirable, subject to the following constraints: First, r_2 and r_3 cannot be chosen simultaneously because they are connected by an "incompatible" edge. Second, r_1, r_2, and r_4 cannot be chosen together, as r_1 and r_4 are reachable via "compatible on the same coalition" edges. By evaluating all possible subsets of rules and comparing their total values, we obtain two optimal feasible rule sets: $\{r_1, r_2\}$ and $\{r_1, r_3, r_4\}$, corresponding to the coalition structures $\{\{a_1, a_2, a_3, a_5\}, \{a_4\}\}$ and $\{a_1, a_3, a_4\}, \{a_2, a_5\}$, respectively, with a social welfare of 5.

When all rules have positive values, the solver developed by Ohta et al. [11] only needs to select as many rules as possible because satisfying more rules never decreases the social welfare. However, when negative values are involved, extra constraints are required to specify when negative value rules should be selected, otherwise, the solver would simply ignore these rules because selecting them decreases the social welfare. This would lead to incorrect results since in some cases, choosing some negative value rules is the precondition of satisfying more positive value rules. For example, given two rules $r_1 : a_1 \wedge a_2 \wedge \neg a_4 \rightarrow -1$ and $r_2 : a_1 \wedge a_2 \wedge a_3 \wedge \neg a_4 \rightarrow 4$, to obtain the value of r_2, the rule r_1 must be included in the feasible rule set, even if it has a negative value.

To address the CSG problem with negative value rules, Ueda et al. [12] introduced zero-valued dummy rules for each negative value rule. A negative value rule is incompatible with all of its dummy rules. The generation rule of dummy rules will be introduced in Definition 8.7. Additionally, Ueda et al. [12] extended Theorem 8.1 by incorporating a constraint that describes when negative value rules must be chosen, as detailed in Theorem 8.2. For the complete proof, see [12].

Theorem 8.2 Ueda et al. [12]. A negative value rule applies to a coalition in coalition structure *CS* if and only if none of its dummy rules apply to any coalition in *CS*.

8.3 Encoding Rule Relations into WPM

Inspired by the direct encoding methods from Ohta et al. [11], Ueda et al. [12], and Liao et al. [21, 22] developed a rule relation-based WPM (RWPM) encoding, showcasing the scalability of the CSG problem with WPM solvers. This section presents the core concepts of RWPM encoding.

First, consider a special case that a rule contains only one agent. In this case, the rule is always selected because any coalition including that single agent will naturally satisfy the rule. To simplify, for the rest of this chapter, we assume that each rule involves at least two agents, with at least one being a positive literal.

8.3.1 Encoding Positive Value Rules

Encoding a positive value rule into WPM formulas is straightforward. Let $R = \{r_1, \dots, r_n\}$ denote a set of positive value rules. For each rule r_i, a new Boolean variable B_i $(i = 1, \dots, n)$ is introduced, where $B_i = 1$ indicates r_i is in a feasible set of rules. Define $\mathscr{B}$ as the set of all such Boolean variables, i.e., $\mathscr{B} = \{B_1, \dots, B_n\}$.

To handle rule relations described in Theorem 8.1, ensuring reachability is crucial. For instance, if rule r_i is compatible on the same coalition with r_j, and r_j is compatible on the same coalition with r_k, then logically, r_i should also be compatible on the same coalition with r_k. To address such reachability, we introduce a Boolean variable $S_{i,j}$ for each pair of rules r_i and r_j, where $1 \leq i < j \leq n$. Here, $S_{i,j} = 1$ signifies that r_i and r_j are reachable via "compatible on the same coalition," edges in the feasible rule set. Enforcing this reachability requires a set of essential clauses known as transitive laws, including:

- $\neg S_{i,j} \vee \neg S_{j,k} \vee S_{i,k}$,
- $\neg S_{i,j} \vee \neg S_{i,k} \vee S_{j,k}$,
- $\neg S_{i,k} \vee \neg S_{j,k} \vee S_{i,j}$,

where $1 \leq i < j < k \leq n$. The number of hard clauses for representing the transitive laws over R is $n \cdot (n-1) \cdot (n-2)/2$. These laws ensure that the compatibility relationships between rules propagate transitively through the set of rules R, contributing to the logical consistency of the RWPM encoding approach.

Definition 8.5 ***(WPM encoding of positive value rules)*** Let R be a set of positive value rules. For each positive value rule $r_i \in R$ (i.e., $r_i : I_i \to w_i$, $w_i > 0$), a weighted soft clause (B_i, w_i) is introduced. The possible relations between two rules r_i and r_j are encoded as following hard clauses:

(1) If r_i and r_j are "compatible on the same coalition," three hard clauses are generated: $\neg B_i \vee \neg B_j \vee S_{i,j}$, $\neg S_{i,j} \vee B_i$, and $\neg S_{i,j} \vee B_j$.
(2) If r_i and r_j are "compatible on different coalitions," one hard clause is generated: $\neg B_i \vee \neg B_j \vee \neg S_{i,j}$.
(3) If r_i and r_j are "incompatible," one hard clause is generated: $\neg B_i \vee \neg B_j$.
(4) If r_i and r_j are "independent," then no clause is generated.

Definition 8.5 encodes the rule relations outlined in Definition 8.4 into WPM formulas. In this framework, soft clauses represent the positive value rules, while hard clauses define constraints on these rules to determine their applicability to coalitions in *CS*.

Hereafter, let *Hard*(*R*) represent a collection of hard clauses that encompass transitive laws and those specified by Definition 8.5.

Lemma 8.1 Let R be a set of positive value rules, $R' = \{r'_1, r'_2, \dots, r'_l\} \subseteq R$ be a feasible rule set, and $\mathcal{B}' = \{B'_1, B'_2, \dots, B'_l\}$ be the corresponding Boolean variable set. Then, $\mathcal{B}' \cup Hard(R)$ is satisfiable.

Proof: We first address the hard clauses from Definition 8.5 (1). If $B_i \in \mathcal{B}'$ and $B_j \in \mathcal{B}'$, then r_i and r_j are both in R' and are compatible on the same coalition, so we can set $S_{i,j} = 1$. If either B_i or B_j is not in $\mathcal{B}'$, then we set $S_{i,j} = 0$, as either r_i or r_i is not in R'. Thus, the hard clauses from Definition 8.5 (1) are satisfied.

Next, consider the hard clauses from Definition 8.5 (2). If either B_i or B_j is not in $\mathcal{B}'$, the value of $S_{i,j}$ is unrestricted. If both B_i and B_j are in $\mathcal{B}'$, then r_i and r_j are not reachable according to Theorem 8.1 (2), so we set $S_{i,j} = 0$. Therefore, the hard clauses from Definition 8.5 (2) are satisfied.

Finally, we examine the hard clauses from Definition 8.5 (3). Suppose B_i and B_j are both in $\mathcal{B}'$. By Theorem 8.1 (1), either r_i or r_j is not in R', which contradicts the assumption that both are in R'. Hence, the transitive laws are also satisfied under this assignment.

In summary, $\mathcal{B}' \cup Hard(R)$ is satisfiable, completing the proof.

Lemma 8.2 Let R be a set of positive value rules, and $\mathcal{B}' = \{B'_1, B'_2, \dots, B'_l\}$ be a subset of $\mathcal{B}$. If $\mathcal{B}' \cup Hard(R)$ is satisfiable, then $\{r'_1, r'_2, \dots, r'_l\}$ is a feasible rule set.

Proof: The proof is essentially involves the reverse process of proving Lemma 8.1. Given $\mathcal{B}' \cup Hard(R)$ is satisfiable, the corresponding rule set $R' = \{r'_1, r'_2, \dots, r'_l\}$ includes no pair of rules that are marked as "incompatible" according to Definition 8.5 (3). Furthermore, if any two rules in R' are compatible on different coalitions, they are not reachable via the "compatible on the same coalition" relation as per Definition 8.5 (2). Consequently, R' is a feasible rule set.

The above two lemmas establish that if R' is a feasible rule set, then $\mathcal{B}' \cup Hard(R)$ is satisfiable, and vice versa. Therefore, the following theorem holds.

Theorem 8.3 The encoding defined in Definition 8.5, along with the transitive laws, enables a WPM solver to produce the correct results for the CSG problem, which consists of a set of positive value rules.

Proof: We need to verify that the soft clause encoded in Definition 8.5 is correct, i.e., the result output by the WPM solver equals the value of the feasible rule set R'.

If r_i is in a feasible rule set, then the corresponding B_i is set to 1, and both of $I_i \to w_i$ and (B_i, w_i) yield a payoff of w_i. Otherwise, if r_i is not in a feasible rule set, then $B_i = 0$ and the corresponding payoffs of both formulas are 0. Given a feasible rule set $R' = \{r'_1, r'_2, \dots, r'_l\}$ and the corresponding Boolean variable set $B' = \{B'_1, B'_2, \dots, B'_l\}$, Lemma 8.1 implies that $B' \cup Hard(R)$ is satisfiable. This means $\forall B'_i \in B'$, (B'_i, w_i) is satisfied, leading the WPM solver to produce a result of $\sum_{i=1}^{l} w_i$, which equals the value of the feasible rule set R'. Alternatively, we can also prove that the sum of weights of the satisfied soft clauses equals the value of the feasible rule set R' by Lemma 8.2.

Example 8.9 Let there be four agents a_1, a_2, a_3, a_4 and four positive value rules:

$$r_1 : a_1 \wedge a_2 \to 2, \qquad r_2 : a_2 \wedge a_3 \wedge \neg a_4 \to 2,$$
$$r_3 : a_1 \wedge a_4 \to 1, \qquad r_4 : a_3 \wedge \neg a_2 \to 3$$

We introduce four soft clauses $(B_1, 2)$, $(B_2, 2)$, $(B_3, 1)$, and $(B_4, 3)$, respectively. Their rule relations are captured by the following clauses:

- r_1 and r_2 are compatible on the same coalition: $\neg B_1 \vee \neg B_2 \vee S_{1,2}$, $\neg S_{1,2} \vee B_1$, $\neg S_{1,2} \vee B_2$.
- r_1 and r_3 are compatible on the same coalition: $\neg B_1 \vee \neg B_3 \vee S_{1,3}$, $\neg S_{1,3} \vee B_1$, $\neg S_{1,3} \vee B_3$.
- r_1 and r_4 are compatible on different coalitions: $\neg B_1 \vee \neg B_4 \vee \neg S_{1,4}$.
- r_2 and r_3 are compatible on different coalitions: $\neg B_2 \vee \neg B_3 \vee \neg S_{2,3}$.
- r_2 and r_4 are incompatible: $\neg B_2 \vee \neg B_4$.
- r_3 and r_4 are independent: no clause is generated.

The four rules yield a total of 12 hard clauses for describing the transitive laws. Altogether, to solve this problem, 10 variables and 25 clauses are generated. The complete file in WPM input format is provided in Appendix D.

All the clauses are connected by logical conjunction and input to a WPM solver. The solver then outputs the optimized social welfare and the corresponding truth assignment, which allows for the derivation of the optimal feasible rule set. The assigned truth values of all the Boolean variables are listed in Figure 8.4.

Based on the variable assignments $B_1 = 1$, $B_2 = 0$, $B_3 = 1$, and $B_4 = 1$, we can immediately conclude that the optimal feasible rule set is $\{r_1, r_3, r_4\}$. This

$B_1 = 1$	$B_3 = 1$	$S_{1,3} = 1$	$S_{2,4} = 0$
$B_2 = 0$	$B_4 = 1$	$S_{1,4} = 0$	$S_{3,4} = 0$
	$S_{1,2} = 0$	$S_{2,3} = 0$	

Figure 8.4 Assignment of all Boolean variables in Example 8.9 (Source: Xiaojuan Liao and Miyuki Koshimura/original).

corresponds to the coalition structure $\{\{a_1, a_2, a_4\}, \{a_3\}\}$, with a maximized social welfare of 6.

8.3.2 Encoding Positive Value Embedded Rules

To make an embedded rule $er \in ER$ (i.e., $er : I_0|I_1, \dots, I_k \rightarrow w_{er}$) apply to (C, CS), each $I_1, \dots, I_k$ must apply to some coalition other than C. To explicitly represent this implicit constraint, Ueda et al. [12] developed an *explicit form* for each embedded rule. This was achieved by adding each positive literal in I_0 to the negative literals of each $I_1, \dots, I_k$, and adding each positive literal in $I_1, \dots, I_k$ to the negative literals of I_0. However, transforming an embedded rule into its explicit form can significantly increase the number of agents. Specifically, let $pos_0, \dots, pos_k$ be the number of positive literals in $I_0, \dots, I_k$, respectively. In the worst case, the number of agents that need to be added in er is $\left(k \cdot pos_0 + \sum_{x=1}^{k} pos_x\right)$. This increase in agents would be prohibitive when the number of embedded rules grows, especially in approaches that rely on relationships between pairs of agents. Therefore, it is crucial to reduce the number of agents in the explicit form.

The explicit form of $er : I_0|I_1, \dots, I_k \rightarrow w_{er}$ is reformed as follows. For each I_i ($i \in \{1, \dots, k\}$), we first check whether $P_0 \cap N_i \neq \emptyset$ or $P_i \cap N_0 \neq \emptyset$ holds. If either condition is true, it means I_0 and I_i cannot apply to the same coalition naturally, and no further action is needed. Otherwise, we add one agent in P_i to N_0. This ensures that the coalition I_i applies to (suppose C) contains at least one agent in N_0, making it impossible for I_0 to apply to C. This explicit form guarantees that I_0 and I_i ($i \in \{1, \dots, k\}$) cannot apply to the same coalition. This method significantly reduces the number of agents that need to be added to *er*.

Example 8.10 For an embedded rule $er : a_1 \wedge a_2 | a_4 \wedge \neg a_3 \rightarrow 1$, the explicit form developed in [12] for *er* is $a_1 \wedge a_2 \wedge \neg a_4 | a_4 \wedge \neg a_3 \wedge \neg a_1 \wedge \neg a_2 \rightarrow 1$, while in the method introduced in this subsection, the explicit form is simplified as $a_1 \wedge a_2 \wedge \neg a_4 | a_4 \wedge \neg a_3 \rightarrow 1$.

The explicit form has already indicated that for each I_i ($i = 1, \dots, k$), it is impossible that I_0 and I_i apply to the same coalition. Therefore, to make *er* apply to some embedded coalition, we only need to have each of $I_0, I_1, \dots, I_k$ apply to some coalition in CS, without any other constraints. For simplicity, we will assume for the remainder of this chapter that each embedded rule is expressed in the reformed explicit form.

For a positive value embedded rule $er : I_0|I_1, \dots, I_k \rightarrow w_{er}$ $(w_{er} > 0)$, we define Boolean variables $B_{er}, B_{er}^0, \dots, B_{er}^k$ corresponding to $er, r_{er}^0, r_{er}^1, \dots, r_{er}^k$, where each r_{er}^i ($i = 0, \dots, k$) is an auxiliary rule $r_{er}^i : I_i \rightarrow 0$. Here, $B_{er} = 1$ indicates that *er* applies to some embedded coalition in CS, while $B_{er}^i = 1$ means that r_{er}^i applies to some coalition in CS. Whether an embedded rule applies to some embedded coalition is determined solely by whether its auxiliary rules are feasible. Therefore, constraints on rule relations and transitive laws for reachability can be specified using only the auxiliary rules, rather than the embedded rules.

Definition 8.6 ***(WPM encoding for positive value embedded rules)*** Let ER be a set of positive value embedded rules in the explicit form. For each positive value embedded rule $er \in ER$ (i.e., $er : I_0|I_1, \dots, I_k \rightarrow w_{er}(w_{er} > 0)$), the following clauses are introduced:

- **One weighted soft clause:** (B_{er}, w_{er}), and
- **$(k+1)$ hard clauses:** $\neg B_{er} \vee B^i_{er}$, for $i = 0, \dots, k$.

The possible relations between any two auxiliary rules and the transitive laws among auxiliary rules are defined in the same way as introduced in Section 8.3.1.

Theorem 8.4 The encoding given in Definition 8.6 ensures that a WPM solver output correct results for the CSG problem, which consists of a set of positive value embedded rules.

Proof: The encoding of rule relations with transitive laws was elaborated in Section 8.3.1. Here, we need to verify the correctness of the encoding on the positive value embedded rule in Definition 8.6.

If all Boolean variables $B^0_{er}, \dots, B^k_{er}$ are true, then no hard clauses restrict B_{er}. Consequently, the WPM solver will favor $B_{er} = 1$ to maximize the payoff w_{er}.

Conversely, if any of $B^0_{er}, \dots, B^k_{er}$ are false, then the corresponding hard clause cannot be satisfied unless $B_{er} = 0$. In this case, the payoff is 0.

In both cases, the WPM solver correctly computes the social welfare.

Example 8.11 Let there be four agents a_1, a_2, a_3, a_4 and four rules with positive values:

$$r_1 : a_1 \wedge a_2 \rightarrow 2, \qquad r_2 : a_3 \wedge \neg a_4 \rightarrow 1,$$
$$r_3 : a_4 \wedge \neg a_1 \rightarrow 1, \qquad r_4 : a_1 \wedge a_2 \wedge \neg a_4 | a_4 \wedge \neg a_3 \rightarrow 1$$

We introduce four weighted soft clauses $(B_1, 2)$, $(B_2, 1)$, $(B_3, 1)$ and $(B_4, 1)$ corresponding to rules r_1, r_2, r_3, and r_4, respectively. The embedded rule r_4 generates two auxiliary rules: $r_5 : a_1 \wedge a_2 \wedge \neg a_4 \rightarrow 0$, $r_6 : a_4 \wedge \neg a_3 \rightarrow 0$. The constraints linking r_4 with its auxiliary rules are specified by two hard clauses: $\neg B_4 \vee B_5$, $\neg B_4 \vee B_6$, as specified in Definition 8.6. The rule relations are captured by the following clauses:

- r_1 and r_5 are compatible on the same coalition: $\neg B_1 \vee \neg B_5 \vee S_{1,5}$, $\neg S_{1,5} \vee B_1$, $\neg S_{1,5} \vee B_5$.
- r_3 and r_6 are compatible on the same coalition: $\neg B_3 \vee \neg B_6 \vee S_{3,6}$, $\neg S_{3,6} \vee B_3$, $\neg S_{3,6} \vee B_6$.

$B_1 = 1$	$B_5 = 1$	$S_{1,4} = 0$	$S_{2,5} = 0$	$S_{3,6} = 1$
$B_2 = 1$	$B_6 = 1$	$S_{1,5} = 1$	$S_{2,6} = 0$	$S_{4,5} = 0$
$B_3 = 1$	$S_{1,2} = 0$	$S_{1,6} = 0$	$S_{3,4} = 0$	$S_{4,6} = 0$
$B_4 = 1$	$S_{1,3} = 0$	$S_{2,3} = 0$	$S_{3,5} = 0$	$S_{5,6} = 0$
		$S_{2,4} = 0$		

Figure 8.5 Assignment of all Boolean variables in Example 8.11 (Source: Xiaojuan Liao and Miyuki Koshimura/original).

- r_1 and r_3 are compatible on different coalitions: $\neg B_1 \vee \neg B_3 \vee \neg S_{1,3}$.
- r_2 and r_3 are compatible on different coalitions: $\neg B_2 \vee \neg B_3 \vee \neg S_{2,3}$.
- r_2 and r_6 are compatible on different coalitions: $\neg B_2 \vee \neg B_6 \vee \neg S_{2,6}$.
- r_3 and r_5 are compatible on different coalitions: $\neg B_3 \vee \neg B_5 \vee \neg S_{3,5}$.
- r_5 and r_6 are compatible on different coalitions: $\neg B_5 \vee \neg B_6 \vee \neg S_{5,6}$.
- r_1 and r_2, r_1 and r_6, and r_2 and r_5 are independent: no clause is generated.

The four rules yield a total of 12 hard clauses for describing the transitive laws. Altogether, to solve this problem, 21 variables and 47 clauses were generated. For the complete WPM input format file, refer to Appendix D.

All the clauses are connected by logical conjunction and input to a WPM solver. The solver then outputs the optimized social welfare and the corresponding truth assignment, which allows for the derivation of the optimal feasible rule set. The assigned truth values of all the Boolean variables are listed in Figure 8.5.

Based on the variable assignments $B_1 = 1$, $B_2 = 1$, $B_3 = 1$, and $B_4 = 1$, we can immediately conclude that the optimal feasible rule set is $\{r_1, r_2, r_3, r_4\}$. This corresponds to the coalition structure $\{\{a_1, a_2\}, \{a_3\}, \{a_4\}\}$, with a maximized social welfare of 5.

8.3.3 Encoding Negative Value Rules

Encoding negative value rules into WPM formulas is more complex than translating positive value rules. First, WPM requires clause weights to be positive, and most WPM solvers can only process clauses with weights greater than zero. Thus, converting negative values to positive ones is essential. Additionally, constraints associated with a negative value rule and its corresponding dummy rules [12] must be encoded into propositional Boolean formulas that WPM solvers can handle.

Definition 8.7 ***(WPM encoding of negative value rules)*** Let R be a set of negative value rules. For each negative value rule $r_x \in R$ (i.e., $r_x : I_x \rightarrow w_x$, where $w_x < 0$), a weighted soft clause $(\neg B_x, -w_x)$ is introduced. Suppose $I_x = a_1 \wedge a_2 \wedge \cdots \wedge a_k \wedge \neg a_{k+1} \wedge \neg a_{k+2} \wedge \cdots \wedge \neg a_m$. The rule r_x generates $m - 1$ dummy rules, which are of two types:

i) $r_x^i : a_1 \wedge \neg a_{i+1} \rightarrow 0$ where $1 \leq i \leq k-1$,
ii) $r_x^j : a_1 \wedge a_{j+1} \rightarrow 0$ where $k \leq j \leq m-1$,

where r_x^i represents the ith dummy rule derived from r_x.

The constraints on r_x and its dummy rules are expressed through the following hard clauses:

(1) $B_x \vee \bigvee_{i \in \{1,\ldots,m-1\}} B_x^i$, and
(2) $\neg B_x^i \vee \neg B_x$, for $i = 1, \ldots, m-1$,

where B_x^i is a Boolean variable.

Intuitively, $B_x^i = 1$ if r_x^i applies to a coalition C in CS and $B_x^i = 0$, otherwise. The final social welfare, after encoding, increases by $(-W_{neg})$ compared to the original one, where W_{neg} is the sum of values of the negative value rules.

From Definition 8.7, it is evident that for a negative value rule r_x, each of its dummy rules r_x^i is always incompatible with r_x. This incompatibility arises because r_x^i and r_x both contain the positive literal a_1, but r_x^i either has a positive literal of r_x as its negative literal or a negative literal of r_x as its positive literal. Therefore, a negative value rule is selected only if none of its dummy rules are selected, as described in Theorem 8.2. The rigorous proof is provided in [12].

Theorem 8.5 The encoding provided by Definitions 8.5 and 8.7, along with the transitive laws, ensures that a WPM solver outputs the correct results for the CSG problem, which includes a set of negative value rules.

Proof: Given a negative value rule r_x, if r_x is in a feasible rule set, i.e., $B_x = 1$, then the payoffs are 0 and w_x for $(\neg B_x, -w_x)$ and $I_x \rightarrow w_x$, respectively. Conversely, if r_x is not in a feasible rule set, i.e., $B_x = 0$, the payoffs are $-w_x$ for $(\neg B_x, -w_x)$ and 0 for $I_x \rightarrow w_x$. Clearly, the payoff of $(\neg B_x, -w_x)$ is constantly $(-w_x)$ more than that of $I_x \rightarrow w_x$. For a coalition game with n negative value rules, denoted by $r_1, r_2, \ldots, r_n$, the social welfare after encoding negative value rules is $(-W_{neg})$ more than the original, where $W_{neg} = \sum_{x=1}^{n} w_x$.

The hard clauses introduced by Definition 8.7 (1) and (2) are derived from Theorem 8.2. We now prove that this encoding is correct.

Let R' be a feasible rule set and B' be the corresponding Boolean variable set. Let $D(r_x) = \{r_x^1, r_x^2, \ldots, r_x^{m-1}\}$ be the set of dummy rules created by a negative value rule r_x, and $B_{D(r_x)} = \{B_x^1, B_x^2, \ldots, B_x^{m-1}\}$ be the set of corresponding Boolean variables.

Consider the hard clause introduced by Definition 8.7 (1). If for all $r_x^i \in D(r_x)$, $r_x^i \notin R'$ (i.e., $B_x^i \notin B'$), then B_x must be in B' according to this hard clause, meaning that r_x must be in R'. This aligns with the "if" part of Theorem 8.2.

Next, consider the hard clauses from Definition 8.7 (2). If $r_x \in R'$ (i.e., $B_x \in \mathcal{B}'$), then B_x^i must not in $\mathcal{B}'$ for all $B_x^i \in B_{D(r_x)}$, meaning that none of the dummy rules r_x^i are in R'. This is consistent with the "only if" part of Theorem 8.2.

With this encoding, the CSG problem is transformed into one involving only positive value rules. The remaining steps follow those outlined in Definition 8.5. Detailed proofs for this part are provided in Lemmas 8.1, 8.2, and Theorem 8.3.

It is important to note that the construction of dummy rules ensures that each negative value rule is incompatible with all of its dummy rules. This "incompatibility" is effectively captured by Definition 8.5. Furthermore, if there exists a dummy rule $r_x^i \in D(r_x)$ such that $r_x^i \notin R'$ (i.e., $B_x^i \notin \mathcal{B}'$), the value of B_x is not constrained by Definition 8.7 (2). In this case, a WPM solver will naturally prefer $B_x = 0$ to maximize the total weight of satisfied soft clauses. Therefore, in practice, hard clauses in Definition 8.7 (2) is redundant and can be omitted safely.

After encoding the negative value rules and the constraints on dummy rules, the remaining steps in solving the problem are the same as handling positive value rules, as detailed in Section 8.3.1. This involves identifying the rule relations between each pair of rules and generating the corresponding hard clauses for these relations, as well as hard clauses for transitive laws. Note that when identifying rule relations and representing the transitivity of these relations, it is essential to include the newly generated dummy rules in the analysis. Specifically, let R_d be the set of dummy rules. The various types of rule relations encoded in Definition 8.4 should be identified over the combined set $R \cup R_d$. Additionally, the number of hard clauses needed to represent the transitive laws should be updated to $(n + n_d)(n + n_d - 1)(n + n_d - 2)/2$, where n_d is the cardinality of R_d.

Example 8.12 Let there be four agents a_1, a_2, a_3, a_4 and three rules:

$$r_1 : a_1 \wedge a_2 \wedge \neg a_4 \rightarrow -1, \quad r_2 : a_1 \wedge a_2 \wedge a_3 \wedge \neg a_4 \rightarrow 4, \; r_3 : a_2 \wedge a_4 \rightarrow 2$$

We introduce three weighted soft clauses $(\neg B_1, 1)$, $(B_2, 4)$, $(B_3, 2)$, respectively for r_1, r_2, and r_3. r_1 generates two dummy rules: $r_4 : a_1 \wedge \neg a_2 \rightarrow 0$ and $r_5 : a_1 \wedge a_4 \rightarrow 0$. For r_4 and r_5, Boolean variables B_4 and B_5 are introduced. The constraint imposed on r_1 and its dummy rules is captured by $B_1 \vee B_4 \vee B_5$, and the rule relations are captured by the following clauses:

- r_1 and r_2 are compatible on the same coalition: $\neg B_1 \vee \neg B_2 \vee S_{1,2}$, $\neg S_{1,2} \vee B_1$, $\neg S_{1,2} \vee B_2$.
- r_3 and r_5 are compatible on the same coalitions: $\neg B_3 \vee \neg B_5 \vee S_{3,5}$. $\neg S_{3,5} \vee B_3$, $\neg S_{3,5} \vee B_5$.
- r_4 and r_5 are compatible on the same coalition: $\neg B_4 \vee \neg B_5 \vee S_{4,5}$, $\neg S_{4,5} \vee B_4$, $\neg S_{4,5} \vee B_5$.

$B_1 = 1$	$B_4 = 0$	$S_{1,3} = 0$	$S_{2,3} = 0$	$S_{3,4} = 0$
$B_2 = 1$	$B_5 = 0$	$S_{1,4} = 0$	$S_{2,4} = 0$	$S_{3,5} = 0$
$B_3 = 0$	$S_{1,2} = 1$	$S_{1,5} = 0$	$S_{2,5} = 0$	$S_{4,5} = 0$

Figure 8.6 Assignment of all Boolean variables in Example 8.12 (Source: Xiaojuan Liao and Miyuki Koshimura/original).

- r_3 and r_4 are compatible on different coalitions: $\neg B_3 \lor \neg B_4 \lor \neg S_{3,4}$.
- r_1 and r_3 are incompatible: $\neg B_1 \lor \neg B_3$.
- r_1 and r_4 are incompatible: $\neg B_1 \lor \neg B_4$.
- r_1 and r_5 are incompatible: $\neg B_1 \lor \neg B_5$.
- r_2 and r_3 are incompatible: $\neg B_2 \lor \neg B_3$.
- r_2 and r_4 are incompatible: $\neg B_2 \lor \neg B_4$.
- r_2 and r_5 are incompatible: $\neg B_2 \lor \neg B_5$.

The five rules yield a total of 30 hard clauses for describing the transitive laws. Altogether, to solve this problem, 15 variables and 50 clauses are generated. For the complete file in WPM input format, readers may refer to Appendix F.

All the clauses are connected by logical conjunction and input to a WPM solver. The solver then outputs the optimized social welfare and the corresponding truth assignment, which allows for the derivation of the optimal feasible rule set. The assigned truth values of all the Boolean variables are listed in Figure 8.6.

The solver outputs a solution value of 4. The actual social welfare is determined by subtracting $(-W_{neg}) = 1$ from the solver's result, calculated as $4 - 1 = 3$. Given the variable assignments $B_1 = 1$, $B_2 = 1$, and $B_3 = 0$, we can conclude that the optimal feasible rule set is $\{r_1, r_2\}$, which corresponds to the coalition structure $\{\{a_1, a_2, a_3\}, \{a_4\}\}$.

8.3.4 Encoding Negative Value Embedded Rules

The WPM encoding of negative value embedded rules is similar to that introduced in Section 8.3.3, both derived from Theorem 8.2.

Definition 8.8 ***(WPM encoding for negative value embedded rules)*** Let ER be a set of negative value embedded rules in the explicit form. For each negative value embedded rule $er \in ER$ ($er : I_0|I_1, \dots, I_k \rightarrow w_{er}$, $w_{er} < 0$), where $I_i = a_1 \land a_2 \land \cdots \land a_l \land \neg a_{l+1} \land \cdots \land \neg a_m$, $i = 0, \dots, k$. I_i generates the following two types of dummy rules:

- $r^x_{er,i}: a_1 \land \neg a_{x+1} \rightarrow 0$, where $1 \le x \le l - 1$,
- $r^y_{er,i}: a_1 \land a_{y+1} \rightarrow 0$, where $l \le y \le m - 1$,

where $r^x_{er,i}$ denotes the xth dummy rules of a condition I_i in er.

For the negative value embedded rule *er*, the following clauses are introduced:

- **One soft clause:** $(\neg B_{er}, -w_{er})$,
- **One hard clause:** $B_{er} \vee \bigvee_{\substack{i \in \{0,\dots,k\} \\ x \in \{1,\dots,m\}}} B^x_{er,i}$,

where $B^x_{er,i}$ is a Boolean variable corresponding to $r^x_{er,i}$.

After this encoding, the result output by a WPM solver is $(-W_{neg})$ larger than the social welfare of the CSG problem in PFGs, where W_{neg} is the sum of values of negative value embedded rules.

Theorem 8.6 The encoding provided by Definitions 8.6 and 8.8 ensures that a WPM solver produces the correct results for the CSG problem, which consists of a set of negative value embedded rules.

Proof: The correctness of Definition 8.6 has been proved in Theorem 8.4. Here, we demonstrate the accuracy of the encoding in Definition 8.8.

Consider a negative value embedded rule *er*. If *er* does not apply to any coalition in *CS* (i.e., $B_{er} = 0$), the payoff from the soft clause $(\neg B_{er}, -w_{er})$ will be $-w_{er}$, while the payoff from *er* will be 0. Conversely, if *er* applies to some embedded coalition in *CS* (i.e., $B_{er} = 1$), the soft clause $(\neg B_{er}, -w_{er})$ yields a payoff of 0, and *er* yields w_{er}. In both cases, the payoff from $(\neg B_{er}, -w_{er})$ is constantly $(-w_{er})$ more than that from *er*. Consequently, for a coalition game of n negative value embedded rules, the social welfare after encoding is $(-W_{neg})$ greater than the original, where W_{neg} is the sum of values of these rules.

Next, we prove the hard clause introduced by Definition 8.8 is accurately encoded according to Theorem 8.2. If for all $i \in \{0,\dots,k\}$ and $x \in \{1,\dots,m\}$, $B^x_{er,i} = 0$, then B_{er} must be set to 1 to satisfy the hard clause. This coincides with the "if" part of Theorem 8.2. On the other hand, if any $B^x_{er,i}$ is 1, B_{er} is not constrained by the hard clause. In such a case, a WPM solver will favor $B_{er} = 0$ because it allows gaining a payoff of $(-w_{er})$. This is consistent with the converse-negative proposition of the "only if" condition in Theorem 8.2.

Example 8.13 Let there be four agents a_1, a_2, a_3, a_4 and four rules:

$$r_1 : a_1 \wedge a_2 \rightarrow 2, \qquad r_2 : a_3 \wedge \neg a_4 \rightarrow 1,$$
$$r_3 : a_4 \wedge \neg a_1 \rightarrow -1, \qquad r_4 : a_1 \wedge a_2 \wedge \neg a_4 | a_4 \wedge \neg a_3 \rightarrow -1$$

We introduce four weighted soft clauses $(B_1, 2)$, $(B_2, 1)$, $(\neg B_3, 1)$, and $(\neg B_4, 1)$, respectively. The rule r_3 generates a dummy rule $r_5 : a_1 \wedge a_4 \rightarrow 0$, with the constraint specified by $B_3 \vee B_5$. The rule r_4 generates three dummy rules: $r_6 : a_1 \wedge$

$B_1 = 1$	$S_{1,3} = 0$	$S_{2,6} = 0$	$S_{4,6} = 0$
$B_2 = 1$	$S_{1,4} = 0$	$S_{2,7} = 0$	$S_{4,7} = 0$
$B_3 = 0$	$S_{1,5} = 1$	$S_{2,8} = 0$	$S_{4,8} = 0$
$B_4 = 0$	$S_{1,6} = 0$	$S_{3,4} = 0$	$S_{5,6} = 0$
$B_5 = 1$	$S_{1,7} = 1$	$S_{3,5} = 0$	$S_{5,7} = 1$
$B_6 = 0$	$S_{1,8} = 0$	$S_{3,6} = 0$	$S_{5,8} = 0$
$B_7 = 1$	$S_{2,3} = 0$	$S_{3,7} = 0$	$S_{6,7} = 1$
$B_8 = 0$	$S_{2,4} = 0$	$S_{3,8} = 0$	$S_{6,8} = 0$
$S_{1,2} = 0$	$S_{2,5} = 0$	$S_{4,5} = 0$	$S_{7,8} = 0$

Figure 8.7 Assignment of all Boolean variables in Example 8.13 (Source: Xiaojuan Liao and Miyuki Koshimura/original).

$\neg a_2 \rightarrow 0, r_7 : a_1 \wedge a_4 \rightarrow 0, r_8 : a_4 \wedge a_3 \rightarrow 0$, with the constraint given by $B_4 \vee B_6 \vee B_7 \vee B_8$.

The subsequent steps involve handling positive value rules as outlined in Section 8.3.1. This includes identifying the rule relations between each pair of rules, generating corresponding hard clauses for these relations, and creating hard clauses to represent transitive laws.

For this problem, 36 variables and 143 clauses are generated. These include 4 weighted soft clauses representing rules, 2 hard clauses for the negative value (embedded) rules, 32 hard clauses for rule constraints, and 105 hard clauses for transitive laws. The complete WPM input file is available in Appendix F.

All clauses are logically conjoined and provided to a WPM solver. The solver then outputs the optimized social welfare and the corresponding truth assignment, which allows for the derivation of the optimal feasible rule set. The assigned truth values for all Boolean variables are detailed in Figure 8.7.

The solver outputs a solution value of 5. The actual social welfare is recovered by subtracting $(-W_{neg}) = 1 + 1 = 2$ from the solver's result, calculated as $5 - 2 = 3$. Given the variable assignments $B_1 = 1, B_2 = 1, B_3 = 0$ and $B_4 = 0$, we can conclude that the optimal feasible rule set is $\{r_1, r_2\}$, which corresponds to the coalition structure $\{\{a_1, a_2, a_4\}, \{a_3\}\}$.

8.4 Performance Evaluation

This section evaluates the performance of the rule relation-based WPM (RWPM) approach and the direct encoding algorithm [11, 12] for solving CSG problems in CFGs and PFGs. RWPM, using Sat4j (version 2.2.3) [26], was executed on a Core i5-2540 2.6 GHz processor with 8 GB RAM and outperformed other WPM solvers [22]. The direct encoding algorithm, utilizing ILOG CPLEX (version 11.2), ran on a Xeon E5540 2.53 GHz processor with 24 GB RAM.

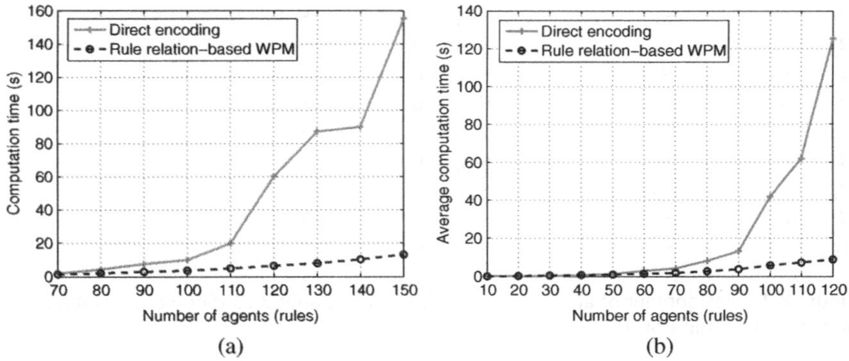

Figure 8.8 Average computation time of RWPM and direct encoding [11, 12]. (a) Computation time for MC-nets (Source: Xiaojuan Liao et al. (2012)/with permission of 24th International Conference on Tools with Artificial Intelligence). (b) Computation time for embedded MC-nets (Source: Xiaojuan Liao et al. (2014)/with permission of IEICE Transactions on Information and Systems).

The instance generation method is detailed in [12]. Throughout the experiments, the number of rules is set as the number of agents (i.e., *#rules* = *#agents*), with approximately 20% of the rules having negative values. Each condition includes around 2.2 agents. For MC-nets (CSG in CFGs), 20% of the literals in a coalition are negative. For embedded MC-nets (CSG in PFGs), the percentage is 15%. For each fixed number of rules, 100 problem instances are generated.

Figure 8.8 illustrates RWPM's superior performance compared to the direct encoding algorithm, especially for larger *#rules*. For MC-nets, as seen in Figure 8.8a, the computation time of the direct encoding algorithm escalates rapidly, exceeding 150 seconds when *#rules* reaches 150. In contrast, RWPM's computation time increases more gradually, reaching approximately 13 seconds for *#rules* = 150. Similarly, for embedded MC-nets, the direct encoding algorithm's computation time sharply rises beyond 90 *#rules*, reaching about 125 seconds for *#rules* = 120. Meanwhile, RWPM maintains a smooth increase in computation time, taking only about 10 seconds for the same instances, as illustrated in Figure 8.8b.

8.5 Chapter Summary

Coalition structure generation (CSG) is a main research topic in coalition games. The CSG problem is typically set in two contexts: Characteristic function games (CFGs), where the value of a coalition is independent of others, and partition function games (PFGs), where the value of a coalition is influenced by the formation of other coalitions.

The challenge of solving the CSG problem stems from the exponential growth in possible solutions and the substantial representation size for enumerating coalition payoffs: $O(2^n)$ for CSG in CFGs and $O(n^n)$ for CSG in PFGs. To address this, MC-nets and embedded MC-nets were introduced as concise representation schemes for CFGs and PFGs, respectively. These schemes represent the CSG problem as a set of rules in the form of *condition* → *value*, encapsulating all possible mappings between coalitions and their values. Thus, solving the CSG problem involves maximizing the sum of values of rules that satisfy certain conditions.

This chapter introduces a rule relation-based WPM (RWPM) method to leverage WPM solvers for addressing the CSG problem in CFGs and PFGs, represented by MC-nets and embedded MC-nets, respectively. Although RWPM is derived from other optimization algorithms, it significantly enhances the efficiency of solving the CSG problem. This marks the first application of WPM encoding to solve the CSG problem.

Chapter 9, an alternative WPM encoding technique will be introduced, along with a performance evaluation of different WPM encodings.

References

1 T. Sandholm and V. R. Lesser. Coalitions among computationally bounded agents. *Artificial Intelligence*, 94:99–137, 1997.

2 V. D. Dang, R. K. Dash, A. Rogers, and N. R. Jennings. Overlapping coalition formation for efficient data fusion in multi sensor network. In *Proceedings of 21st National Conference on Artificial Intelligence and the 18th Innovative Applications of Artificial Intelligence Conference*, 2006.

3 M. Tsvetovat and K. Sycara. Customer coalitions in the electronic marketplace. In *Proceedings of the 4th International Conference on Autonomous Agents*, 2000.

4 M. J. Osborne and A. Rubinstein. *A Course in Game Theory*. The MIT Press, 1994.

5 T. Sandholm, K. Larson, M. Andersson, O. Shehory, and F. Tohme. Coalition structure generation with worst case guarantees. *Artificial Intelligence*, 111:209–238, 1999.

6 E. Catilina and R. Feinberg. Market power and incentives to form research consortia. *Review of Industrial Organization*, 28:129–144, 2008.

7 J. Plasmans, J. Engwerda, B. Aarle, B. Bartolomeo, and T. Michalak. *Dynamic Modeling of Monetary and Fiscal Cooperation Among Nations*. Springer, 2006.

8 S. Ieong and Y. Shoham. Marginal contribution nets: A compact representation scheme for coalitional games. In *Proceedings of 6th ACM Conference on Electronic Commerce*, 2005.

9 T. Rahwan, T. Michalak, M. Wooldridge, and N. R. Jennings. Anytime coalition structure generation in multi-agent systems with positive or negative externalities. *Artificial Intelligence*, 186:95–122, 2012.

10 T. Michalak, D. Marciniak, M. Szamotulski, T. Rahwan, M. Wooldridge, P. McBurney, and N. Jennings. A logic-based representation for coalitional games with externalities. In *Proceedings of 9th International Conference on Autonomous Agents and Multiagent Systems*, 2010.

11 N. Ohta, V. Conitzer, R. Ichimura, Y. Sakurai, A. Iwasaki, and M. Yokoo. Coalition structure generation utilizing compact characteristic function representation. In *Proceedings of 15th International Conference on Principles and Practice of Constraint Programming (CP '09)*, 2009.

12 S. Ueda, T. Hasegawa, N. Hashimoto, N. Ohta, A. Iwasaki, and M. Yokoo. Handling negative value rules in MC-net-based coalition structure generation. In *Proceedings of 11th International Joint Conference on Autonomous Agents and Multiagent Systems*, 2012.

13 D Yun Yeh. A dynamic programming approach to the complete set partitioning problem. *BIT Numerical Mathematics*, 26(4):467–474, 1986.

14 Talal Rahwan, Tomasz P Michalak, and Nicholas R Jennings. A hybrid algorithm for coalition structure generation. In *AAAI*, pages 1443–1449, 2012.

15 Amir Hussin and Shaheen Fatima. Heuristic methods for optimal coalition structure generation. In *Multi-Agent Systems and Agreement Technologies*, pages 124–139. Springer, 2016.

16 Juan Pablo Contreras, Paul Bosch, Mauricio Varas, and Franco Basso. A new genetic algorithm encoding for coalition structure generation problems. *Mathematical Problems in Engineering*, 2020(1):1203248, 2020.

17 Jianguo Jiang, Na Xia, Meibin Qi, and Chunmei Mu. An ant colony algorithm based multi-task coalition serial generation algorithm. *Acta Electronica Sinica*, 33(12):2178–2182, 2005.

18 Guofu Zhang, Renzhi Yang, Zhaopin Su, Feng Yue, Yuqi Fan, Meibin Qi, and Jianguo Jiang. Using binary particle swarm optimization to search for maximal successful coalition. *Applied Intelligence*, 42(2):195–209, 2015.

19 T. Rahwan, T. P. Michalak, E. Elkind, P. Faliszewski, J. Sroka, M. Wooldridge, and N.R. Jennings. Constrained coalition formation. In *Proceedings of the 25th AAAI Conference on Artificial Intelligence*, 2011.

20 S. Ueda, M. Kitaki, A. Iwasaki, and M. Yokoo. Concise characteristic function representations in coalitional games based on agent types. In *Proceedings of 22nd International Joint Conference on Artificial Intelligence*, 2011.

21 Xiaojuan Liao, Miyuki Koshimura, Hiroshi Fujita, and Ryuzo Hasegawa. Solving the coalition structure generation problem with MaxSAT. In *IEEE 24th International Conference on Tools with Artificial Intelligence, ICTAI 2012*, Athens, Greece, November 7-9, 2012, pages 910–915. IEEE Computer Society, 2012.

22 Xiaojuan Liao, Miyuki Koshimura, Hiroshi Fujita, and Ryuzo Hasegawa. MaxSAT encoding for MC-net-based coalition structure generation problem with externalities. *IEICE Transactions on Information & Systems*, 97-D(7):1781–1789, 2014.

23 Xiaojuan Liao, Miyuki Koshimura, Hiroshi Fujita, and Ryuzo Hasegawa. Extending MaxSAT to solve the coalition structure generation problem with externalities based on agent relations. *IEICE Transactions on Information & Systems*, 97-D(7):1812–1821, 2014.

24 Xiaojuan Liao and Miyuki Koshimura. A comparative analysis and improvement of MaxSAT encodings for coalition structure generation under MC-nets. *Journal of Logic and Computation*, 29(6):913–931, 2019.

25 Xiaojuan Liao, Miyuki Koshimura, Kazuki Nomoto, Suguru Ueda, Yuko Sakurai, and Makoto Yokoo. Improved WPM encoding for coalition structure generation under mc-nets. *Constraints*, 24(1):25–55, 2019.

26 D.L. Berre and A. Parrain. The SAT4J library, release 2.2. *Journal on Satisfiability, Boolean Modeling and Computation*, 7:59–64, 2010.

9

Agent Relation-Based Weighted Partial MaxSAT (AWPM) Encoding

After reading this chapter, you should be able to:

- Extend weighted partial MaxSAT (WPM) to handle negative weights.
- Apply WPM to solve the coalition structure generation (CSG) problem based on agent relations.
- Evaluate different WPM encodings in a rational way.

9.1 Extended Weighted Partial MaxSAT

A *weighted clause* is defined as a pair (C, w), where C represents a clause and w denotes its weight, which is a positive integer or infinity. When the weight is infinity, the clause is termed *hard*; otherwise, it is referred to as *soft*. A *weighted partial MaxSAT* (WPM) instance comprises a multiset of weighted clauses $\phi = (C_1, w_1), \ldots, (C_m, w_m), (C_{m+1}, \infty), \ldots, (C_{m+m'}, \infty)$, with the first m clauses being soft and the last m' clauses being hard.

A *truth assignment* for ϕ maps each variable in ϕ to either 0 or 1. The *benefit* of a truth assignment I on ϕ, denoted *benefit*(ϕ, I), is the sum of the weights of the soft clauses satisfied by I if I satisfies all hard clauses. If I fails to satisfy any hard clause, *benefit*$(\phi, I) = -\infty$. A WPM instance ϕ is considered *satisfiable* if there exists a truth assignment that satisfies all hard clauses; otherwise, it is *unsatisfiable*. The goal of the WPM problem for a given instance ϕ is to find an optimal assignment I that maximizes *benefit*(ϕ, I), ensuring that *benefit*$(\phi, I) \geq$ *benefit*(ϕ, I') for any other assignment I'.

In WPM, it is standard practice to restrict the weights of soft clauses to positive values[1] because the primary goal of MaxSAT is to *maximize the number*

1 Currently, the Python toolkit PySAT can handle WCNF formulas with negative weights, although this is not consistent with the WCNF format description.

Applied Satisfiability: Cryptography, Scheduling, and Coalitional Games, First Edition. Xiaojuan Liao and Miyuki Koshimura.

of satisfied clauses. However, some problems, such as the coalition structure generation (CSG) problem, naturally involve formulas with both positive and negative weights, as the value of each coalition can be either positive or negative. Additionally, current WPM solvers are typically limited to handling formulas in conjunction normal form (CNF), which is a conjunction of clauses. However, many real-world combinatorial optimization problems do not naturally present themselves as CNF formulas.

This section introduces an extended version of WPM that can manage propositional Boolean formulas with both positive and negative weights and explores the relationship between the solutions of standard WPM and this extended version.

9.1.1 EWPM-to-WPM Transformation

An extended weighted partial MaxSAT (EWPM) instance is an extension of WPM and consists of a multiset of weighted formulas, denoted as:

$$\phi = (F_1, w_1), \ldots, (F_m, w_m), (F_{m+1}, \infty), \ldots, (F_{m+m'}, \infty)$$

where the first m formulas are soft and the last m' formulas are hard. A weighted formula is a pair (F, w), where F is a propositional Boolean formula and w is a nonzero integer or infinity. A soft formula is termed *positive* if its weight is positive, and *negative* if its weight is negative.

For an EWPM instance ϕ^E and a truth assignment I^E, the *benefit* of I^E on ϕ^E, denoted as *benefit*(ϕ^E, I^E), is the sum of the weights of the soft formulas satisfied by I^E, provided that I^E satisfies all the hard formulas. If I^E does not satisfy all the hard formulas, then *benefit*$(\phi^E, I^E) = -\infty$. The instance ϕ^E is *satisfiable* if there exists a truth assignment that satisfies all the hard formulas; otherwise, it is *unsatisfiable*. The EWPM problem for an instance ϕ^E is to find an optimal assignment I^E that maximizes *benefit*(ϕ^E, I^E), i.e., *benefit*$(\phi^E, I^E) \geq$ *benefit*$(\phi^E, I^{E'})$ for any arbitrary assignment $I^{E'}$.

Given that all hard clauses are satisfied, a natural interpretation of a soft formula (F, w) is that we gain w dollars when F is true. Under this interpretation, a negative soft formula $(F, -20)$ means that we lose 20 dollars when F is true. According to the following Definition 9.1, we transform $(F, -20)$ to $(\neg b, 20)$ where b is logically equivalent to F. $(\neg b, 20)$ indicates that we gain 20 dollars when b is true, i.e., F is false.

This transformation might seem odd because the original soft formula $(F, -20)$ implies that neither gain nor loss is obtained when F is false. To address this, consider that we pay a deposit of $-w$ dollars for a negative soft formula (F, w) (where $w < 0$) and have $-w$ dollars refunded when F is false.

For example, with $(F, -20)$, we first pay 20 dollars as a deposit, then examine the truth value of F. If F is false, we get a 20-dollar refund, resulting in a final payoff of

0, aligning with the meaning of $(F, -20)$. If F is true, we owe 20 dollars according to $(F, -20)$. Since the deposit of 20 dollars has already been paid, this amount is considered in the calculation. The total deposit is the negative sum of all negative weights, denoted as negative W_{neg} in Lemma 9.1, 9.2, and Theorem 9.1.

Definition 9.1 ***(EWPM-to-WPM Transformation)*** Let $\phi^E = \{(F_1, w_1), \dots, (F_m, w_m), (F_{m+1}, \infty), \dots, (F_{m+m'}, \infty)\}$ be an EWPM instance where w_i is nonzero integer and (F_i, w_i) is a soft formula ($1 \leq i \leq m$). The EWPM-to-WPM transformation consists of two steps:

(1) For each soft formula (F_i, w_i), a new variable b_i is introduced. The soft formula is then transformed into a soft clause:

(i) $\begin{matrix} (b_i, w_i) & \textit{if } w_i > 0 \\ (\neg b_i, -w_i) & \textit{if } w_i < 0 \end{matrix}$

and two hard formulas ensuring that F_i and b_i are logically equivalent:

(ii) $(F_i \rightarrow b_i, \infty)$

(iii) $(b_i \rightarrow F_i, \infty)$

(2) Transform hard formulas into hard clauses using a satisfiability-preserving CNF transformation.

In step (1), a new Boolean variable b_i is introduced to be logically equivalent to F_i, allowing the soft formula (F_i, w_i) to be converted into a soft clause (b_i, w_i) or $(\neg b_i, -w_i)$, as shown in (i). The logical equivalence between b_i and F_i is established by (ii) and (iii). Step (2) encodes the hard formulas (ii) and (iii) into a set of clauses using satisfiability-preserving CNF transformations. In this step, any satisfiability-preserving CNF transformations can be applied [1].

Lemma 9.1 Let ϕ^E be a satisfiable EWPM instance, ϕ be a WPM instance derived from ϕ^E using the EWPM-to-WPM transformation, and W_{neg} be the sum of all negative weights in ϕ^E.

If I^E is a truth assignment of ϕ^E that satisfies all the hard formulas, then there exists a truth assignment I of ϕ such that I satisfies all the hard clauses in ϕ and $benefit(\phi^E, I^E) = benefit(\phi, I) + W_{neg}$.

Proof: Assume that I^E satisfies p negative soft formulas and falsifies q negative soft formulas, i.e., I^E satisfies (F_i^s, w_i^s) and falsifies (F_j^f, w_j^f) where $w_i^s < 0$ and $w_j^f < 0$ for $1 \leq i \leq p$ and $1 \leq j \leq q$. Then, $benefit(\phi^E, I^E) = W_{pos}^{I^E} + \sum_{i=1}^{p} w_i^s$, where $W_{pos}^{I^E}$ is the sum of the weights of positive soft clauses satisfied by I^E. Note that $\sum_{i=1}^{p} w_i^s + \sum_{j=1}^{q} w_j^f = W_{neg}$.

Construct a truth assignment I for ϕ by extending I^E to satisfy all the hard clauses in ϕ as follow: If I^E satisfies a soft formula (F_i, w_i), let I satisfy the new variable b_i

introduced by the transformation; otherwise, let I falsify b_i. Evidently, I satisfies the hard formulas (ii) and (iii) in the transformation, thus satisfying all the hard clauses in ϕ because the satisfiability-preserving CNF transformation was used in step (2) of the transformation.

From p negative soft formulas (F_i^s, w_i^s), we get p soft clauses $(\neg b_i^s, -w_i^s)$, and from q negative soft formulas (F_j^s, w_j^s), we get q soft clauses $(\neg b_j^f, -w_j^f)$ by the transformation, where b_i^s and b_j^f are new variables introduced. I falsifies p soft clauses $(\neg b_i^s, -w_i^s)$ and satisfies q soft clauses $(\neg b_j^f, -w_j^f)$. Therefore, *benefit*$(\phi, I) = W_{pos}^I + \sum_{j=1}^q (-w_j^f)$, where W_{pos}^I is the sum of the weights of positive soft clauses satisfied by I. Here, since the transformation does not alter the satisfiability of positive soft formulas, $W_{pos}^I = W_{pos}^{I^E}$ holds. Therefore, *benefit*$(\phi, I) = W_{pos}^{I^E} + \sum_{j=1}^q (-w_j^f)$.

The difference between *benefit*(ϕ^E, I^E) and *benefit*(ϕ, I) is *benefit*$(\phi^E, I^E) -$ *benefit*$(\phi, I) = \sum_{i=1}^p w_i^s - \sum_{j=1}^q (-w_j^f) = \sum_{i=1}^p w_i^s + \sum_{j=1}^q w_j^f = W_{neg}$. Consequently, *benefit*$(\phi^E, I^E) =$ *benefit*$(\phi, I) + W_{neg}$.

Lemma 9.2 Consider ϕ^E, ϕ, and W_{neg} as described in Lemma 9.1. If I is a truth assignment of ϕ and satisfies all the hard clauses in ϕ, then there exists a truth assignment I^E of ϕ^E such that I^E satisfies all the hard formulas in ϕ^E and *benefit*$(\phi^E, I^E) =$ *benefit*$(\phi, I) + W_{neg}$.

Proof: Construct a truth assignment I^E by restricting I to the variables in ϕ^E. Following a similar approach to the proof of Lemma 9.1, we conclude that I^E satisfies all the hard formulas in ϕ^E and the benefits are appropriately adjusted.

Theorem 9.1 Let ϕ^E be a satisfiable EWPM instance and ϕ be a WPM instance obtained from ϕ^E through the EWPM-to-WPM transformation. Then, ϕ is satisfiable. Moreover, if I_{sol}^E is an EWPM solution for ϕ^E and I_{sol} is a WPM solution for ϕ, then *benefit*$(\phi^E, I_{sol}^E) =$ *benefit*$(\phi, I_{sol}) + W_{neg}$, where W_{neg} is the total of all negative weights in ϕ^E.

Proof: According to Lemma 9.1, there exists a truth assignment I' for ϕ such that *benefit*$(\phi^E, I_{sol}^E) =$ *benefit*$(\phi, I') + W_{neg}$. Similarly, according to Lemma 9.2, there exists a truth assignment I'^E for ϕ^E such that *benefit*$(\phi^E, I'^E) =$ *benefit*$(\phi, I_{sol}) + W_{neg}$.

Since *benefit*(ϕ^E, I_{sol}^E) and *benefit*(ϕ, I_{sol}) are maximal benefits, the inequalities *benefit*$(\phi^E, I_{sol}^E) \geq$ *benefit*(ϕ^E, I'^E) and *benefit*$(\phi, I_{sol}) \geq$ *benefit*(ϕ, I') hold.

These inequalities along with the two equalities above, imply that *benefit*(ϕ, I_{sol}) $=$ *benefit*(ϕ, I') and *benefit*$(\phi^E, I_{sol}^E) =$ *benefit*$(\phi, I_{sol}) + W_{neg}$.

The EWPM-to-WPM transformation involves encoding a formula F_i into a logically equivalent Boolean variable b_i through a series of hard clauses that ensure their logical equivalence. This allows the soft formula (F_i, w_i) to be converted into a soft clause (b_i, w_i) or $(\neg b_i, -w_i)$. This transformation keeps the assignment quality intact, merely replacing F_i with b_i. Since F_i and b_i are logically equivalent. Given their logical equivalence, the truth assignment I^E_{sol} remains consistent with I_{sol}.

In the context of Theorem 9.1, the total deposit is $-W_{neg}$ dollars. Consequently, the statement ($benefit(\phi^E, I^E_{sol}) = benefit(\phi, I_{sol}) + W_{neg}$) in Theorem 9.1 indicates that the maximal benefit from ϕ^E equals the maximum benefit from ϕ minus the total deposit.

9.1.2 Redundancy in Transformation

The hard formula (ii) for a positive soft formula and the hard formula (iii) for a negative soft formula are unnecessary for solving the EWPM problem. In other words, Theorem 9.1 remains valid even without these hard formulas.

Proposition 9.1 Consider a satisfiable EWPM instance ϕ^E and a WPM instance ϕ derived from ϕ^E via the EWPM-to-WPM transformation. Let ϕ^- represent a multiset of weighted clauses as follows:

(1) soft clauses in ϕ generated from step (1)-(i),
(2) hard clauses in ϕ generated from step (1)-(iii) if (F_i, w_i) is a positive formula,
(3) hard clauses in ϕ, generated from step (1)-(ii) if (F_i, w_i) is a negative formula,
(4) hard clauses in ϕ generated from step (2).

In essence, ϕ^- is obtained from ϕ by removing the hard clauses from (ii) for positive soft formulas and (iii) for negative soft formulas.

If I_{sol} is a WPM solution for ϕ and I^-_{sol} is a WPM solution for ϕ^-, then $benefit(\phi, I_{sol}) = benefit(\phi^-, I^-_{sol})$.

Proof: Since ϕ is a superset of ϕ^-, and the soft clauses in ϕ are identical to those in ϕ^-, it follows that I_{sol} is also a truth assignment for ϕ^- and satisfies all the hard clauses in ϕ^-. Hence, $benefit(\phi, I_{sol}) \leq benefit(\phi^-, I^-_{sol})$ because I^-_{sol} provides the maximal benefit for ϕ^-.

To prove $benefit(\phi, I_{sol}) \geq benefit(\phi^-, I^-_{sol})$, we need to show that I^-_{sol} also satisfies all the hard clauses in ϕ.

Consider a positive soft formula $(F_i, w_i)(w_i > 0)$. For this formula, the corresponding soft clause (b_i, w_i) is created. If I^-_{sol} falsifies the hard formula (ii) $(F_i \rightarrow b_i, \infty)$, it means that I^-_{sol} satisfies F_i but falsifies b_i. In this case, a truth assignment, that agrees to I^-_{sol} except for the assignment to b_i, gives a benefit of ϕ^-, which is w_i greater than $benefit(\phi^-, I^-_{sol})$. This contradicts the assumption that $benefit(\phi^-, I^-_{sol})$ provides the maximum benefit. Therefore, I^-_{sol} must satisfy the hard formula (ii). Consequently, I^-_{sol} satisfies the hard clauses from (ii), which are not included in ϕ^-.

Similarly, we can show that I^-_{sol} satisfies the hard clauses derived from hard formula (iii) for a negative soft formula. Therefore, I^-_{sol} satisfies all the hard clauses in ϕ.

Proposition 9.1 relieves the EWPM-to-WPM transformation of redundant clauses. If the formula to be handled is positive, (ii) is skipped, otherwise, (iii) can be omitted safely. In the experiments in Section 9.3, we also omit these redundant clauses.

9.1.3 MinSAT Extension

Let us consider MinSAT, an alternative to MaxSAT. MinSAT involves finding a truth assignment that satisfies all the hard clauses while minimizing the sum of the weights of the satisfied soft clauses. In the context of EWPM, we can think of it as "minimizing" the sum of absolute values of weights of satisfied negative soft formulas. Therefore, it is possible to approach solving MinSAT using the EWPM framework. Let $\phi^{Min} = \{(C_1, w_1), \dots, (C_m, w_m), (C_{m+1}, \infty), \dots, (C_{m+m'}, \infty)\}$ be a satisfiable weighted partial MinSAT instance. We make an EWPM instance ϕ^{Max} by negating the weights in ϕ^{Min}: $\phi^{Max} = \{(C_1, -w_1), \dots, (C_m, -w_m), (C_{m+1}, \infty), \dots, (C_{m+m'}, \infty)\}$

A truth assignment that satisfies all the hard clauses in ϕ^{Min} will also satisfy all the hard clauses in ϕ^{Max}, and vice versa. It is important to note that $\mathit{benefit}(\phi^{Max}, I) = -\mathit{benefit}(\phi^{Min}, I)$ where I is a truth assignment. This means that as $\mathit{benefit}(\phi^{Min}, I)$ decreases, $\mathit{benefit}(\phi^{Max}, I)$ increases correspondingly.

Therefore, a MinSAT solution for ϕ^{Min}, which minimizes the benefit of ϕ^{Min}, corresponds to a MaxSAT solution for ϕ^{Max}, which maximizes the benefit of ϕ^{Max}. Conversely, a MaxSAT solution for ϕ^{Max} corresponds to a MinSAT solution for ϕ^{Min}.

In summary, ϕ^{Min} and ϕ^{Max} share the same solution. Thus, EWPM can be viewed not only as an extension of MaxSAT but also as an extension of MinSAT, effectively integrating both problems.

9.2 Encoding Agent Relations into WPM

Liao et al. [2] introduced a rule relation-based WPM (RWPM) encoding to address the CSG problem. Building on this work, they later proposed an agent relation-based WPM (AWPM) encoding, transforming the CSG problem into a series of EWPM formulas. This approach significantly enhances the efficiency of solving the CSG problem under general parameter configurations [3]. This section focuses on AWPM and demonstrates its capability in encoding agent relations.

Before diving into this chapter, we strongly recommend that readers first review Chapter 8 to familiarize themselves with the concepts of the CSG problem, MC-nets, and embedded MC-nets. Similar to Chapter 8, this chapter assumes each rule involves at least two agents, with one being a positive literal.

9.2.1 Agent Relation

Let $\mathscr{A}_r$ denote the set of agents in rule r, $\mathscr{A}_r^-$ the set of negative literals in r, and $\mathscr{A}_r^+$ the set of positive literals in r. We introduce a Boolean variable $\mathscr{C}_{i,j}$ for any pair of agents a_i and a_j where $i < j$. The variable is defined as follow: $\mathscr{C}_{i,j} = 1$ if a_i and a_j are in the same coalition, and $\mathscr{C}_{i,j} = 0$ otherwise.

Given a rule r of m agents, we designate an agent a_p as a standard agent if p is the smallest index among all positive literals in r. Formally, $a_p \in \mathscr{A}_r^+$ and for any other positive literal a_l in $\mathscr{A}_r^+$, $p < l$. With a_p identified, the agent relations within r can be represented by the relationships between a_p and each other agent a_k where $a_k \in \mathscr{A}_r \setminus \{a_p\}$. The Boolean variable $\mathscr{C}_{p,k}$ (when $p < k$) or $\mathscr{C}_{k,p}$ (when $p > k$) denotes the agent relation between a_p and a_k. Intuitively, $\mathscr{C}_{k,p}$ is equal to 1 if $a_k \in \mathscr{A}_r^+$, and equal to 0 if $a_k \in \mathscr{A}_r^-$.

An MC-net [4] provides a compact representation of the CSG problem, expressing it as a set of rules in the form of *condition* → *value*. According to Definition 8.1, the condition of a rule can be formally written as:

$$\{a_1 \wedge a_2 \wedge \cdots \wedge a_l \wedge \neg a_{l+1} \wedge \cdots \wedge \neg a_m\}.$$

In this context, the standard agent is a_1. We represent the condition of a rule by using the agent relations between the standard agent a_1 and other agents in $\mathscr{A}_r$, as described in Definition 9.2.

Definition 9.2 ***(Agent relation-based MC-nets)*** An agent relation-based MC-net consists of a set of rules R. Each rule $r_i \in R$ is syntactically represented as $\mathcal{S}i \to w_i$, where $\mathcal{S}i$ is a conjunction of agent relations, formally expressed by $\mathscr{C}_{1,2} \wedge \cdots \wedge \mathscr{C}_{1,l} \wedge \neg\mathscr{C}_{1,l+1} \wedge \cdots \wedge \neg\mathscr{C}_{1,m}$. A rule r_i is said to apply to a coalition C if $\mathcal{S}i$ holds true, i.e., $\{\mathscr{C}_{1,k}\}_{k=2}^{l}$ are all true, and $\{\mathscr{C}_{1,k}\}_{k=l+1}^{m}$ are all false.

Definition 9.2 can be extended to embedded MC-nets, as shown in Definition 9.3. Each embedded rule is expressed explicitly, as defined in Section 8.3.2, and summarized as follows: For an embedded rule $er : I_0|I_1, \ldots, I_k \to w_{er}$, check whether $P_0 \cap N_i \neq \emptyset$ or $P_i \cap N_0 \neq \emptyset$ holds for each I_i $(i \in \{1, \cdots, k\})$. If either condition is true, no further action is needed. Otherwise, add a positive literal from I_i to the negative literals of I_0, ensuring I_0 does not apply to any coalition that I_i applies to.

Definition 9.3 ***(Agent relation-based embedded MC-nets)*** An agent relation-based embedded MC-net consists of a set of embedded rules ER. Each rule $er \in ER$ is expressed in the form $\mathcal{S}_0 \wedge \mathcal{S}_1 \wedge \cdots \wedge \mathcal{S}_k \rightarrow w_{er}$, where each $\mathcal{S}_i$: $i \in \{0, \dots, k\}$ is a conjunction of agent relations as defined in Definition 9.2. An embedded rule er applies to an embedded coalition (C, CS) if all $\{\mathcal{S}_i\}_{i=0}^k$ are true.

The relation of two agents being in the same coalition is transitive. Specifically, any three agents a_i, a_j, and a_k $(1 \leq i < j < k \leq m)$ satisfy the following transitive laws:

- $\neg\mathcal{C}_{i,j} \vee \neg\mathcal{C}_{j,k} \vee \mathcal{C}_{i,k}$,
- $\neg\mathcal{C}_{i,j} \vee \neg\mathcal{C}_{i,k} \vee \mathcal{C}_{j,k}$,
- $\neg\mathcal{C}_{i,k} \vee \neg\mathcal{C}_{j,k} \vee \mathcal{C}_{i,j}$,

where m is the number of agents. The number of hard clauses required to represent these transitive laws is $m\cdot (m-1)\cdot (m-2)/2$.

Example 9.1 Let there be four agents a_1, a_2, a_3, a_4 and two rules:

- r_1: $a_1 \wedge a_2 \wedge a_3 \wedge \neg a_4 \rightarrow 1$,
- r_2: $a_1 \wedge a_2 \wedge \neg a_4 | a_4 \wedge \neg a_3 \rightarrow 1$.

By Definitions 9.2 and 9.3, the two rules are expressed as follows:

- r_1: $\mathcal{C}_{1,2} \wedge \mathcal{C}_{1,3} \wedge \neg\mathcal{C}_{1,4} \rightarrow 1$,
- r_2: $\mathcal{C}_{1,2} \wedge \neg\mathcal{C}_{1,4} \wedge \neg\mathcal{C}_{3,4} \rightarrow 1$.

The four agents generate a total of 12 hard clauses by the transitive laws.

The transitive laws mentioned earlier can sometimes be refined. Consider a CSG problem with the following three rules:

$$r_1 : a_1 \wedge a_2 \rightarrow 2, \qquad r_2 : a_3 \wedge \neg a_4 \wedge \neg a_5 \rightarrow 1, \qquad r_3 : a_1 \wedge \neg a_6 \rightarrow 1$$

Here, agents a_3, a_4, and a_5 only appear in r_2 and are not related to a_1, a_2, and a_6. Therefore, we can group a_3, a_4, and a_5 together, and combine the remaining agents into another group. These two groups are independent of each other. As a result, the number of clauses needed to represent the transitive laws is calculated within each group individually. In this example, the number of transitive laws is reduced to 6, compared to the original calculation of $6{\cdot}5{\cdot}4/2 = 60$.

By applying the EWPM-to-WPM transformation and utilizing agent relations, we can encode rules in both MC-nets and embedded MC-nets efficiently. The following Sections 9.2.2–9.2.5 will detail the encodings for both positive value (embedded) rules and negative value (embedded) rules.

9.2.2 Encoding Positive Value Rules

In this subsection, we explore the encoding strategies for positive value rules within MC-nets, ensuring that the agent relations are accurately captured and utilized.

Definition 9.4 ***(WPM encoding for positive value rules)*** For a positive value rule $r_i : \mathcal{S}_i \rightarrow w_i$ $(w_i > 0)$, where $\mathcal{S}_i = \mathcal{C}_{1,2} \wedge \cdots \wedge \mathcal{C}_{1,l} \wedge \neg\mathcal{C}_{1,l+1} \wedge \cdots \wedge \neg\mathcal{C}_{1,m}$, we introduce the following clauses, using a new Boolean variable u_i.

(i) a weighted soft clause: (u_i, w_i), and
(ii) $(m-1)$ hard clauses: $\neg u_i \vee \mathcal{C}_{1,2}, \ldots, \neg u_i \vee \mathcal{C}_{1,l}, \neg u_i \vee \neg\mathcal{C}_{1,l+1}, \ldots, \neg u_i \vee \neg\mathcal{C}_{1,m}$.

Theorem 9.2 The encoding provided in Definition 9.4 along with the transitive laws enables a WPM solver to produce the correct results of the CSG problem, which consists of a set of positive value rules.

Proof: Given a coalition structure *CS*, we can create an assignment $\mathcal{A}$ that satisfies all the hard clauses for the transitive laws, agreeing with *CS*. Conversely, if there is an assignment $\mathcal{A}$ that satisfies all hard clauses for the transitive laws, we can derive a coalition structure *CS* that agrees with $\mathcal{A}$. Hence, there is a one-to-one correspondence between a set of coalition structures and the set of assignments that satisfy all hard clauses for the transitive laws.

Definition 9.4 (i) and (ii) correspond to Definition 9.1. (i) and (iii), respectively. Here, (u_i, w_i) is transformed from the positive formula $(\mathcal{S}_i, w_i)$, and the $(m-1)$ hard clauses are derived from the hard formula $(u_i \rightarrow \mathcal{S}_i, \infty)$. According to Theorem 9.1 and Proposition 9.1, Definition 9.1 (i) and (iii) ensure the correctness of EWPM-to-WPM transformation for positive soft formulas. Thus, WPM solvers compute the correct social welfare.

Example 9.2 Let there be four agents a_1, a_2, a_3, a_4 and four positive value rules:

$$r_1 : a_1 \wedge a_2 \rightarrow 2, \qquad r_2 : a_2 \wedge a_3 \wedge \neg a_4 \rightarrow 2,$$
$$r_3 : a_1 \wedge a_4 \rightarrow 1, \qquad r_4 : a_3 \wedge \neg a_2 \rightarrow 3$$

In the agent relation-based MC-net, these four rules are expressed as follows:

$$r_1 : \mathcal{C}_{1,2} \rightarrow 2, \qquad r_2 : \mathcal{C}_{2,3} \wedge \neg\mathcal{C}_{2,4} \rightarrow 2,$$
$$r_3 : \mathcal{C}_{1,4} \rightarrow 1, \qquad r_4 : \neg\mathcal{C}_{2,3} \rightarrow 3$$

AWPM generates the following soft and hard clauses for these rules:

- r_1 is encoded into a soft clause $(u_1, 2)$ and a hard clause $\neg u_1 \vee \mathcal{C}_{1,2}$,
- r_2 is encoded into a soft clause $(u_2, 2)$ and two hard clauses $\neg u_2 \vee \mathcal{C}_{2,3}$, $\neg u_2 \vee \neg\mathcal{C}_{2,4}$,

$u_1 = 1$ $u_2 = 0$ $u_3 = 1$ $u_4 = 1$	$C_{1,2} = 1$ $C_{1,3} = 0$ $C_{1,4} = 1$	$C_{2,3} = 0$ $C_{2,4} = 1$ $C_{3,4} = 0$

Figure 9.1 Assignment of all Boolean variables in Example 9.2 (Source: Xiaojuan Liao and Miyuki Koshimura/original).

- r_3 is encoded into a soft clause $(u_3, 1)$ and a hard clause $\neg u_3 \vee \mathscr{C}_{1,4}$,
- r_4 is encoded into a soft clause $(u_4, 3)$ and a hard clause $\neg u_4 \vee \neg\mathscr{C}_{2,3}$,

where u_1, u_2, u_3, and u_4 are newly generated Boolean variables.

The four rules yield a total of 12 hard clauses for describing the transitive laws. Altogether, to solve this problem, 10 variables and 21 clauses are generated. The complete file in WPM input format is provided in Appendix H.

All the clauses are connected by logical conjunction and input to a WPM solver. The solver then outputs the optimized social welfare and the corresponding truth assignment, which allows for the derivation of the optimal feasible rule set. The assigned truth values of all the Boolean variables are listed in Figure 9.1.

Based on the variable assignments $u_1 = 1$, $u_2 = 0$, $u_3 = 1$, and $u_4 = 1$, we can immediately conclude that the optimal feasible rule set is $\{r_1, r_3, r_4\}$. This corresponds to the coalition structure $\{\{a_1, a_2, a_4\}, \{a_3\}\}$, with a maximized social welfare of 6.

9.2.3 Encoding Positive Value Embedded Rules

Building on the concepts introduced in Section 9.2.2, this subsection delves into the encoding for positive value embedded rules based on agent relations.

Definition 9.5 ***(WPM encoding for positive value embedded rules)*** For a positive value embedded rule $er : \mathscr{S}_0 \wedge \mathscr{S}_1 \wedge \cdots \wedge \mathscr{S}_k \rightarrow w_{er}$ $(w_{er} > 0)$, we introduce the following clauses, using a new Boolean variable u_{er}:

(i) a weighted soft clause: (u_{er}, w_{er}), and
(ii) $k + 1$ propositional Boolean formulas that can be expanded into multiple hard clauses: $\neg u_{er} \vee \mathscr{S}_0, \ldots, \neg u_{er} \vee \mathscr{S}_1, \neg u_{er} \vee \mathscr{S}_k$.

Theorem 9.3 The encoding provided in Definition 9.5 along with the transitive laws, enables a WPM solver to produce the correct results of the CSG problem, which consists of a set of positive value embedded rules.

Proof: Given a coalition structure *CS*, we can create an assignment $\mathscr{A}$ that satisfies all the hard clauses for the transitive laws, agreeing with *CS*.

If a positive value embedded rule *er* applies to an embedded coalition in *CS*, then $\mathscr{A}$ satisfies each $\mathscr{S}_i$, and the corresponding soft clause (u_{er}, w_{er}) is not restricted by any hard clauses. In this scenario, WPM solvers prefer $u_{er} = 1$ over $u_{er} = 0$ to gain the payoff w_{er}.

If r_{er} does not apply to any embedded coalition in CS, then $\mathscr{A}$ does not satisfy at least one $\mathcal{S}_i$ unless $u_{er} = 0$. Consequently, the corresponding soft clause (u_{er}, w_{er}) is unsatisfied, resulting in a payoff of 0.

In both cases, WPM solvers correctly compute the social welfare.

Example 9.3 Let there be four agents a_1, a_2, a_3, a_4 and four rules with positive values:

$$r_1 : a_1 \wedge a_2 \rightarrow 2, \qquad r_2 : a_3 \wedge \neg a_4 \rightarrow 1,$$
$$r_3 : a_4 \wedge \neg a_1 \rightarrow 1, \qquad r_4 : a_1 \wedge a_2 \wedge \neg a_4 | a_4 \wedge \neg a_3 \rightarrow 1$$

In the agent relation-based MC-net, these four rules are expressed as follows:

$$r_1 : \mathscr{C}_{1,2} \rightarrow 2, \qquad r_2 : \neg\mathscr{C}_{3,4} \rightarrow 1,$$
$$r_3 : \neg\mathscr{C}_{1,4} \rightarrow 1, \qquad r_4 : \mathscr{C}_{1,2} \wedge \neg\mathscr{C}_{1,4} \wedge \neg\mathscr{C}_{3,4} \rightarrow 1$$

AWPM generates the following soft and hard clauses for these rules:

- r_1 is encoded into a soft clause $(u_1, 2)$ and a hard clause $\neg u_1 \wedge \mathscr{C}_{1,2}$,
- r_2 is encoded into a soft clause $(u_2, 1)$ and a hard clause $\neg u_2 \vee \neg\mathscr{C}_{3,4}$,
- r_3 is encoded into a soft clause $(u_3, 1)$ and a hard clause $\neg u_3 \vee \neg\mathscr{C}_{1,4}$,
- r_4 is encoded into a soft clause $(u_4, 1)$ and three hard clauses: $\neg u_4 \vee \mathscr{C}_{1,2}$, $\neg u_4 \vee \neg\mathscr{C}_{1,4}$, $\neg u_4 \vee \neg\mathscr{C}_{3,4}$,

where u_1, u_2, u_3, and u_4 are newly introduced Boolean variables.

The four rules yield a total of 12 hard clauses for describing the transitive laws. Altogether, to solve this problem, 10 variables and 22 clauses are generated. The complete file in WPM input format is provided in Appendix H.

All the clauses are connected by logical conjunction and input to a WPM solver. The solver then outputs the optimized social welfare and the corresponding truth assignment, which allows for the derivation of the optimal feasible rule set. The assigned truth values of all the Boolean variables are listed in Figure 9.2.

Based on the variable assignments $u_1 = 1$, $u_2 = 1$, $u_3 = 1$, and $u_4 = 1$, we can immediately conclude that the optimal feasible rule set is $\{r_1, r_2, r_3, r_4\}$. This corresponds to the coalition structure $\{\{a_1, a_2\}, \{a_3\}, \{a_4\}\}$, with a maximized social welfare of 4.

$u_1 = 1$	$C_{1,2} = 1$	$C_{2,3} = 0$
$u_2 = 1$	$C_{1,3} = 0$	$C_{2,4} = 0$
$u_3 = 1$	$C_{1,4} = 0$	$C_{3,4} = 0$
$u_4 = 1$		

Figure 9.2 Assignment of all Boolean variables in Example 9.3 (Source: Xiaojuan Liao and Miyuki Koshimura/original).

9.2.4 Encoding Negative Value Rules

In contrast to positive value rules, negative value rules denote contributions that detract from the coalition structure's overall value. Encoding these rules requires a different approach to ensure that their negative impact is correctly represented. This subsection will guide you through the encoding process, demonstrating how to handle negative value rules based on the EWPM-to-WPM transformation.

If a rule is a negative value rule $r_i : \mathcal{S}_i \to w_i$ $(w_i < 0)$, as explained in Section 9.1.1, the EWPM-to-WPM transformation referred to Definition 9.1. (i) and (ii), derives the following encodings.

Definition 9.6 ***(WPM encoding for negative value rules)*** For each negative value rule $r_i : \mathcal{S}_i \to w_i$ $(w_i < 0)$, we introduce the following clauses, using a new Boolean variable u_i.

(i) a weighted soft clause: $\left(\neg u_i, -w_i\right)$, and
(ii) a hard clause: $u_i \vee \neg \mathcal{S}_i$.

The social welfare after encoding is increased by $(-W_{neg})$ compared to the original, where W_{neg} is the sum of negative weights.

Theorem 9.4 The encoding provided in Definition 9.6 along with the transitive laws enables a WPM solver to produce the correct results of the CSG problem, which consists of a set of negative value rules.

Proof: Consider a coalition structure CS. We can construct an assignment $\mathcal{A}$ that satisfies all hard clauses for the transitive laws so as to agree with CS.

Definition 9.6. (i) and (ii) align with Definition 9.1. (i) and (ii), respectively. Based on Theorem 9.1 and Proposition 9.1, these definitions ensure the correctness of the EWPM-to-WPM transformation for negative soft formulas. Consequently, WPM solvers compute the correct social welfare.

Example 9.4 Let there be four agents a_1, a_2, a_3, a_4 and three rules:

$$r_1 : a_1 \wedge a_2 \wedge \neg a_4 \to -1, \quad r_2 : a_1 \wedge a_2 \wedge a_3 \wedge \neg a_4 \to 4, \quad r_3 : a_2 \wedge a_4 \to 2$$

In the agent relation-based MC-net, these three rules are expressed as follows:

$$r_1 : \mathcal{C}_{1,2} \wedge \neg \mathcal{C}_{1,4} \to -1, \qquad r_2 : \mathcal{C}_{1,2} \wedge \mathcal{C}_{1,3} \wedge \neg \mathcal{C}_{1,4} \to 4, \quad r_3 : \mathcal{C}_{2,4} \to 2$$

AWPM generates the following soft and hard clauses for these rules:

- r_1 is encoded into a soft clause $(\neg u_1, 1)$ and a hard clause $u_1 \vee \neg \mathcal{C}_{1,2} \vee \mathcal{C}_{1,4}$,
- r_2 is encoded into a soft clause $(u_2, 4)$ and three hard clauses $\neg u_2 \vee \mathcal{C}_{1,2}$, $\neg u_2 \vee \mathcal{C}_{1,3}$, $\neg u_2 \vee \neg \mathcal{C}_{1,4}$.
- r_3 is encoded into a soft clause $(u_3, 2)$ and a hard clause $\neg u_3 \vee \mathcal{C}_{2,4}$,

where u_1, u_2, and u_3 are newly generated Boolean variables.

$u_1 = 1$ $\quad C_{1,2} = 1$ $\quad C_{2,3} = 1$
$u_2 = 1$ $\quad C_{1,3} = 1$ $\quad C_{2,4} = 0$
$u_3 = 0$ $\quad C_{1,4} = 0$ $\quad C_{3,4} = 0$

Figure 9.3 Assignment of all Boolean variables in Example 9.2 (Source: Xiaojuan Liao and Miyuki Koshimura/original).

The three rules yield a total of 12 hard clauses for describing the transitive laws. Altogether, to solve this problem, 9 variables and 20 clauses are generated. The complete file in WPM input format is provided in Appendix J.

All the clauses are connected by logical conjunction and input to a WPM solver. The solver then outputs the optimized social welfare and the corresponding truth assignment, which allows for the derivation of the optimal feasible rule set. The assigned truth values of all the Boolean variables are listed in Figure 9.3.

The solver outputs a solution value of 4. The actual social welfare is determined by subtracting $(-W_{neg}) = 1$ from the solver's output result, calculated as $4 - 1 = 3$. Based on the variable assignments $u_1 = 1, u_2 = 1$, and $u_3 = 0$, we can conclude that the optimal feasible rule set is $\{r_1, r_2\}$. This corresponds to the coalition structure $\{\{a_1, a_2, a_3\}, \{a_4\}\}$.

9.2.5 Encoding Negative Value Embedded Rules

This subsection covers the encoding strategies for negative value-embedded rules, focusing on preserving their detailed structure and ensuring accurate representation in the transformed framework.

Definition 9.7 ***(WPM encoding for negative value embedded rules)*** For each negative value-embedded rule $er : \mathcal{S}_0 \wedge \mathcal{S}_1 \wedge \cdots \wedge \mathcal{S}_k \rightarrow w_{er}$ $(w_{er} < 0)$, we introduce the following clauses, using a new Boolean variable u_i:

(i) a weighted soft clause: $(\neg u_{er}, -w_{er})$, and
(ii) a hard clause: $u_i \vee \neg\mathcal{S}_0 \vee \neg\mathcal{S}_1 \vee \cdots \vee \neg\mathcal{S}_k$.

After encoding, the social welfare is increased by $(-W_{neg})$, where W_{neg} is the sum of negative weights.

Theorem 9.5 The encoding provided in Definition 9.7 along with the transitive laws enables a WPM solver to produce the correct results of the CSG problem, which consists of a set of negative value-embedded rules.

Proof: Consider a coalition structure CS. We can create an assignment $\mathcal{A}$ that satisfies all hard clauses for the transitive laws to agree with CS.

If a negative value embedded rule r_{er} applies to an embedded coalition in CS, then $\mathcal{A}$ satisfies each $\mathcal{S}_i$. This means $\mathcal{A}$ will not satisfy the corresponding hard clause unless $u_{er} = 1$. Consequently, the soft clause $(\neg u_{er}, -w_{er})$ is unsatisfied, resulting in no payoff from WPM solvers. The actual payoff of r_{er} is w_{er} because it applies to an embedded coalition in CS.

If a negative value rule r_{er} does not apply to any embedded coalition in CS, then $\mathscr{A}$ does not satisfy at least one $\mathcal{S}_i$. This means $\mathscr{A}$ satisfies the corresponding hard clause, and the soft clause $(\neg u_{er}, -w_{er})$ is not restricted by any hard clauses. In this scenario, WPM solvers prefer $u_{er} = 0$ because the positive payoff of $-w_{er}$ is gained. The actual payoff of r_{er} is 0 because it does not apply to any coalition in CS.

In both cases, WPM solvers overestimate the social welfare by $-w_{er}$ for a negative value-embedded rule, resulting in a social welfare value that is $(-W_{neg})$ larger than the original.

Example 9.5 Let there be four agents a_1, a_2, a_3, a_4 and four rules:

$$r_1 : a_1 \wedge a_2 \rightarrow 2, \qquad r_2 : a_3 \wedge \neg a_4 \rightarrow 1,$$
$$r_3 : a_4 \wedge \neg a_1 \rightarrow -1, \qquad r_4 : a_1 \wedge a_2 \wedge \neg a_4 | a_4 \wedge \neg a_3 \rightarrow -1$$

In the agent relation-based MC-net, these four rules are expressed as follows:

$$r_1 : \mathscr{C}_{1,2} \rightarrow 2, \qquad r_2 : \neg\mathscr{C}_{3,4} \rightarrow 1,$$
$$r_3 : \neg\mathscr{C}_{1,4} \rightarrow -1, \qquad r_4 : \mathscr{C}_{1,2} \wedge \neg\mathscr{C}_{1,4} \wedge \neg\mathscr{C}_{3,4} \rightarrow -1$$

AWPM generates the following soft and hard clauses for these rules:

- r_1 is encoded into a soft clause $(u_1, 2)$ and a hard clause $\neg u_1 \wedge \mathscr{C}_{1,2}$,
- r_2 is encoded into a soft clause $(u_2, 1)$ and a hard clause $\neg u_2 \vee \neg\mathscr{C}_{3,4}$,
- r_3 is encoded into a soft clause $(\neg u_3, 1)$ and a hard clause $u_3 \vee \mathscr{C}_{1,4}$,
- r_4 is encoded into a soft clause $(\neg u_4, 1)$ and a hard clause $u_4 \vee \neg\mathscr{C}_{1,2} \vee \mathscr{C}_{1,4} \vee \mathscr{C}_{3,4}$,

where u_1, u_2, u_3, and u_4 are newly introduced Boolean variables.

The four rules yield a total of 12 hard clauses for representing transitive laws. Altogether, to solve this problem, a total of 10 variables and 20 clauses are generated. To access the complete file in WPM format, readers may refer to Appendix J.

All the clauses are connected by logical conjunction and input to a WPM solver. The solver then outputs the optimized social welfare and the corresponding truth assignment, which allows for the derivation of the optimal feasible rule set. The assigned truth values of all the Boolean variables are listed in Figure 9.4.

The solver outputs a solution value of 5. The actual social welfare is determined by subtracting $(-W_{neg}) = 1 + 1 = 2$ from the solver's output result, calculated as $5 - 2 = 3$. Based on the variable assignments $u_1 = 1$, $u_2 = 1$, $u_3 = 0$, and $u_4 = 0$, we can conclude that the optimal feasible rule set is $\{r_1, r_2\}$. This corresponds to the coalition structure $\{\{a_1, a_2, a_4\}, \{a_3\}\}$.

$$u_1 = 1 \qquad u_2 = 1 \qquad u_3 = 0 \qquad u_4 = 0$$
$$C_{1,2} = 1 \qquad C_{1,3} = 0 \qquad C_{1,4} = 1$$
$$C_{2,3} = 0 \qquad C_{2,4} = 1 \qquad C_{3,4} = 0$$

Figure 9.4 Assignment of all Boolean variables in Example 9.5 (Source: Xiaojuan Liao and Miyuki Koshimura/original).

9.3 Performance Evaluation

This section compares the performance of the RWPM approach and the AWPM approach for solving the CSG problem in both CFGs (without externalities) and PFGs (with externalities). The experimental conditions match those described in Chapter 8. All experiments were conducted on a Core i5-2540 2.6 GHz processor with 8GB RAM.

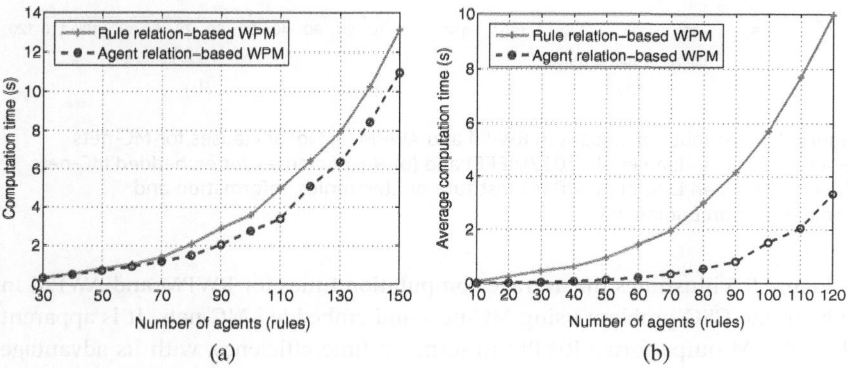

Figure 9.5 Average computation time of RWPM and AWPM (a) Computation time for MC-nets (Source: Xiaojuan Liao et al. (2012)/IEEE) and (b) Computation time for embedded MC-nets (Source: Xiaojuan Liao et al. (2014)/Institute of Electronics, Information and Communication Engineers).

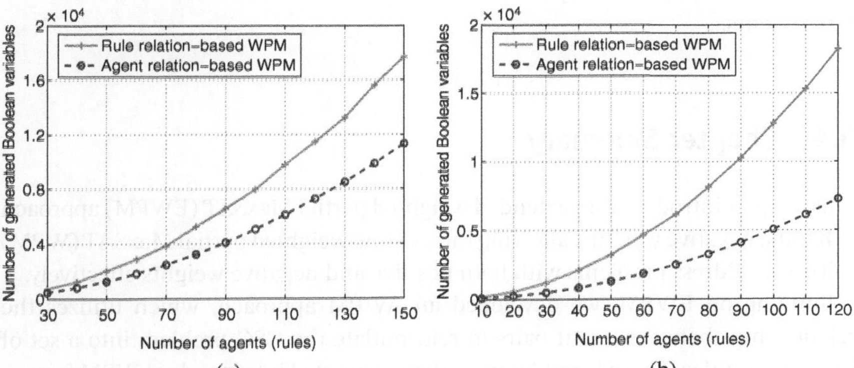

Figure 9.6 Number of Boolean variables in RWPM and AWPM (a) No. of Boolean variables for MC-nets (Source: Xiaojuan Liao et al. (2012)/IEEE) and (b) No. of Boolean variables for embedded MC-nets (Source: Xiaojuan Liao et al. (2014)/Institute of Electronics, Information and Communication Engineers).

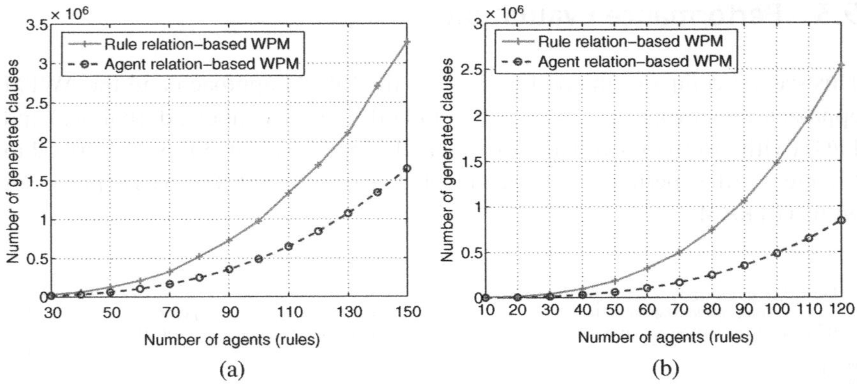

Figure 9.7 Number of clauses in RWPM and AWPM (a) No. of clauses for MC-nets (Source: Xiaojuan Liao et al. (2012)/IEEE) and (b) No. of clauses for embedded MC-nets (Source: Xiaojuan Liao et al. (2014)/Institute of Electronics, Information and Communication Engineers).

Figure 9.5 illustrates the average computation times for RWPM and AWPM in solving the CSG problem using MC-nets and embedded MC-nets. It is apparent that AWPM outperforms RWPM in terms of time efficiency, with its advantage becoming more significant as the number of agents increases.

The increased efficiency of AWPM can be explained by examining the number of Boolean variables and clauses in each approach, as depicted in Figures 9.6 and 9.7. RWPM has a higher number of variables and clauses compared to AWPM. Generally, the computation time increases with the number of variables and clauses, which explains why RWPM takes longer to compute than AWPM in our experiments.

9.4 Chapter Summary

This chapter introduced an extended weighted partial MaxSAT (EWPM) approach to handle negative weights, allowing the existing weighted partial MaxSAT (WPM) solvers to address problems with both positive and negative weights effectively.

Building on EWPM, we developed an AWPM approach, which utilizes the relationships between agent pairs to reformulate the CSG problem into a set of EWPM formulas. These formulas were then converted into standard WPM formulas using the EWPM-to-WPM transformation, enabling the use of existing WPM solvers. Experimental results demonstrated that AWPM outperforms the RWPM introduced in Chapter 8, achieving higher efficiency under common parameter settings. This performance difference can be attributed, in part, to the differing numbers of generated Boolean variables and clauses in the two approaches.

Chapter 10 will explore a curious phenomenon where RWPM and AWPM, despite generating identical numbers of variables and clauses, exhibit markedly different computational efficiencies.

References

1 Armin Biere, Marijn Heule, Hans van Maaren, and Toby Walsh. *Handbook of Satisfiability* - Second Edition. IOS Press, 2021.

2 Xiaojuan Liao, Miyuki Koshimura, Hiroshi Fujita, and Ryuzo Hasegawa. MaxSAT encoding for MC-net-based coalition structure generation problem with externalities. *IEICE Transactions on Information & Systems*, 97-D(7):1781–1789, 2014.

3 Xiaojuan Liao, Miyuki Koshimura, Hiroshi Fujita, and Ryuzo Hasegawa. Extending MaxSAT to solve the coalition structure generation problem with externalities based on agent relations. *IEICE Transactions on Information & Systems*, 97-D(7):1812–1821, 2014.

4 S. Ieong and Y. Shoham. Marginal contribution nets: A compact representation scheme for coalitional games. In *Proceedings of 6th ACM Conference on Electronic Commerce*, 2005.

10

Comparative Analysis and Improvement of RWPM

After reading this chapter, you should be able to:

- Quantitatively compare the rule relation-based WPM (RWPM) and agent relation-based WPM (AWPM) for solving the coalition structure generation (CSG) problem in marginal contribution networks (MC-nets) from both theoretical and experimental views.
- Deduce redundancy when encoding transitive laws in RWPM.
- Understand the idea of improving RWPM with refined transitive laws.

10.1 Motivation

The survey [1] emphasized the significance of weighted partial MaxSAT (WPM) encodings for solving the coalition structure generation (CSG) problem, noting that both rule relation-based WPM (RWPM) and agent relation-based WPM (AWPM) are efficient and faster than other methods [2, 3]. Chapter 9 showed that AWPM outperforms RWPM with specific parameters, especially when the number of agents is between a few dozen and over 100, with each condition having around 2.2 agents. Chapter 9 attributed this superiority to the fewer Boolean variables and clauses in AWPM compared to RWPM, aligning with the understanding that increased variables and clauses generally lead to longer computation times [4, 5].

However, determining if the comparison results will vary with different parameter settings is challenging due to insufficient theoretical support. Liao and Koshimura [6] developed a quantitative method to explore the performance differences between RWPM and AWPM from both theoretical and experimental perspectives. They discovered a special case where RWPM and AWPM generate identical numbers of variables and clauses yet exhibit vastly

Applied Satisfiability: Cryptography, Scheduling, and Coalitional Games,
First Edition. Xiaojuan Liao and Miyuki Koshimura.

different performance. Upon investigation, they refined RWPM by improving its transitive laws.

This chapter focuses on analyzing and improving RWPM for solving the CSG problem in MC-nets. We highly recommend that readers familiarize themselves with RWPM and AWPM encodings in Chapters 8 and 9 before proceeding with this chapter.

10.2 Comparative Study on RWPM and AWPM

In general, adding auxiliary constraints to a constraint satisfiability problem can reduce the solver's search space and improve efficiency. However, in some cases, these constraints can complicate the problem, requiring more inferences to check their impact on variable evaluations. If redundant, these constraints can lead to unnecessary inferences and inferior performance.

In the context of WPM encodings for the CSG problem, Chapter 9 demonstrated that AWPM is more efficient than RWPM under certain parameter settings, generating fewer variables and clauses. This suggests that WPM-solving time increases with the number of variables and clauses. In this section, we will comprehensively compare the numbers of Boolean variables and clauses in these two WPM encodings.

Table 10.1 provides the notations used in our analysis. Table 10.2 presents a summary of the variables and clauses generated by both encodings. We will first explain each item in Table 10.2 and then compare the number of variables and clauses in RWPM and AWPM.

10.2.1 Comparing the Number of Boolean Variables

In general, the Boolean variables generated by RWPM and AWPM for solving the CSG problem in CFGs fall into two categories: (1) variables representing rules and (2) variables indicating the pairwise relations between rules or agents.

For RWPM, the inclusion of dummy rules results in $n + n_d$ variables representing rules, and $C^2_{n+n_d} = (n + n_d)(n + n_d - 1)/2$ variables representing rule relations. On the other hand, AWPM does not introduce auxiliary rules, so the number of variables representing rules is n, and the number of variables representing agent relations is $C^2_m = m(m - 1)/2$. Thus, the total number of Boolean variables generated in RWPM is $(n + n_d) + (n + n_d)(n + n_d - 1)/2$, while for AWPM it is $n + m(m - 1)/2$, respectively.

Theorem 10.1 For any R, given an equal number of rules and agents (*#rules* = *#agents*), RWPM introduces at least as many variables as AWPM.

Table 10.1 Notations and descriptions used in theoretical comparisons between RWPM and AWPM.

Notations	Descriptions
$\lvert S\rvert$	Number of elements in set S
R	A finite set of rules in an MC-net
A	A finite set of agents in an MC-net
n	Total number of rules in R. $n = \lvert R\rvert$
m	Total number of agents in A. $m = \lvert A\rvert$
R_p	A set of positive value rules
R_n	A set of negative value rules. $R_n = R \backslash R_p$
R_d	A set of dummy rules
n_d	Total number of dummy rules. $n_d = \lvert R_d\rvert$
m_i	Number of agents in the condition of r_i, where $r_i \in R$
$\overline{m}$	Average number of agents in a condition
T_r	Number of hard clauses for representing transitive laws in RWPM
T_a	Number of hard clauses for representing transitive laws in AWPM
N_r	Number of hard clauses for specifying constraints on rule relations
N_a	Number of hard clauses for encoding conjunction formulas based on agent relations

Table 10.2 Number of variables and clauses in RWPM and AWPM.

	Variables	Clauses
RWPM	$(n + n_d) + (n + n_d)(n + n_d - 1)/2$	$n + T_r + N_r + \lvert R_n\rvert$
AWPM	$n + m(m - 1)/2$	$n + T_a + N_a$

Specifically, when an MC-net consists solely of positive value rules, both RWPM and AWPM generate the same number of variables.

Proof: If R only contains positive value rules, then $n_d = 0$. Consequently, the number of variables in RWPM is $n + n(n-1)/2$. Given that *#rules* = *#agents*, i.e., $n = m$, the number of variables in RWPM and AWPM is the same.

However, if R contains both positive and negative value rules, additional dummy rules are generated ($n_d > 0$). As a result, RWPM generates more variables than AWPM.

10.2.2 Comparing the Number of Clauses

In RWPM, n weighted soft clauses (B_i, w_i) are used to represent n rules. The satisfiability of these clauses depends on both the hard clauses defined in Definitions 8.5 and 8.7 and the transitive laws of rule relations. For each negative value rule r_x, a hard clause is generated to enforce constraints on r_x and its associated dummy rules, resulting in $|R_n|$ hard clauses for the negative rule set R_n (Definition 8.7). Additionally, T_r hard clauses are used to encode transitive laws of rule relations, where $T_r = 3 \cdot C^3_{n+n_d} = (n + n_d)(n + n_d - 1)(n + n_d - 2)/2$. Let N_r be the number of hard clauses for rule relations over $R \cup R_d$. Therefore, the total number of clauses introduced by RWPM is given by $n + T_r + N_r + |R_n|$.

In AWPM, n weighted soft clauses are introduced for n rules. The encoding also involves generating a set of hard clauses to represent the rules in conjunction form. For each positive value rule $r_i \in R_p$, $(m_i - 1)$ hard clauses are required (as defined in Definition 9.4), while each negative value rule $r_j \in R_n$ needs only one hard clause (as defined in Definition 9.6). Consequently, the total number of hard clauses for encoding conjunction formulas is $N_a = \sum_{r_i \in R_p} (m_i - 1) + |R_n|$. Additionally, T_a hard clauses are used to encode the transitive laws of agent relations, where $T_a = 3 \cdot C^3_m = m\,(m - 1)\,(m - 2)/2$. Overall, AWPM introduces $n + N_a + T_a$ clauses.

To compare the numbers of clauses in RWPM and AWPM, we define each rule r_i as a 2-tuple (I_i, w_i) and the rule set R is represented as $\{(I_i, w_i)\}$. For any R, let R^+ be defined as $R^+ = \{(I_i, |w_i|)\}$. Let C_R and C^+_R denote the number of clauses created by RWPM for R and R^+, respectively. Similarly, C_A and C^+_A represent the clause counts for AWPM with respect to R and R^+.

Lemma 10.1 In RWPM, for any given R, the total number of clauses generated to solve R is always at least as large as that for R^+: $C_R \geq C^+_R$. Specifically, when an MC-net contains solely positive value rules, the number of clauses for R equals that for R^+: $C_R = C^+_R$.

Proof: If R consists solely of positive value rules, then $R = R^+$ and consequently, $C_R = C^+_R$.

If R contains both positive and negative value rules, let T^+_r and N^+_r denote the values corresponding to T_r and N_r in R, but for R^+. Thus, the total number of clauses for R^+ is $C^+_R = n + T^+_r + N^+_r + |R_n|$, where $T^+_r = n\,(n - 1)\,(n - 2)/2$ and $|R_n| = 0$. Since $T_r > T^+_r$ and $N_r > N^+_r$ (because rule relations are identified over $R \cup R_d$ and R_d is empty for R^+), it follows that $C_R > C^+_R$.

Therefore, in both cases, $C_R \geq C^+_R$ holds.

Lemma 10.2 In AWPM, for any given R, the total number of clauses generated to solve R is always at least as large as that for R^+: $C_A \leq C^+_A$. Specifically, when an MC-net consists of solely positive value rules, the number of clauses for R equals that for R^+: $C_A = C^+_A$.

Proof: If R consists solely of positive value rules, then $R = R^+$ and $C_A = C_A^+$.

If R contains both positive and negative value rules, let T_a^+ and N_a^+ represent the values corresponding to T_a and N_a in R, but for R^+. Thus, the total number of clauses for R^+ is $C_A^+ = n + T_a^+ + N_a^+$, where $T_a^+ = m(m-1)(m-2)/2$ and $N_a^+ = \sum_{r_i \in R^+} (m_i - 1)$. Since $T_a^+ = T_a$ and considering that $N_a = \sum_{r_i \in R_p} (m_i - 1) + |R_n|$, where $R_n > 0$ and each rule has at least two agents ($m_i \geq 2$), it follows that $N_a \leq N_a^+$.

Therefore, in both cases, $C_A \leq C_A^+$ holds.

In what follows, we will examine the relationship between C_R^+ and C_A^+. To ease our discussion, we will denote the average number of agents in a condition as $\overline{m}$.

Lemma 10.3 For any R, given an equal number of rules and agents (*#rules = #agents*), the number of clauses generated by RWPM is at least as large as that produced by AWPM when solving R^+, provided that $\overline{m} \leq (m+1)/2$. That is, given $n = m$ and $\overline{m} \leq (m+1)/2$, $C_R^+ \geq C_A^+$ holds. Specifically, when $\overline{m} = (m+1)/2$, then $C_R^+ = C_A^+$.

Proof: As established in Lemmas 10.1 and 10.2, $C_R^+ = n + T_r^+ + N_r^+$ and $C_A^+ = n + T_a^+ + N_a^+$, where $T_r^+ = n(n-1)(n-2)/2$ and $T_a^+ = m(m-1)(m-2)/2$. With $n = m$, it follows that $T_r^+ = T_a^+$, leading to the difference $C_R^+ - C_A^+ = N_r^+ - N_a^+$.

Introducing $\overline{m}$, we approximate $N_a^+ = \sum_{r_i \in R^+} (m_i - 1)$ as $n(\overline{m} - 1)$. The value of N_r^+ depends on the rule relations in R^+. Let P denote the probability that any two rules share one or more agents, indicating nonindependent relations. According to Definition 8.5, each pair of related rules generates at least one hard clause, so $N_r^+ \geq PC_n^2$.

For $\overline{m} \leq m/2$, we have

$$P = 1 - \frac{C_m^{\overline{m}} C_{m-\overline{m}}^{\overline{m}}}{C_m^{\overline{m}} C_m^{\overline{m}}} = 1 - \frac{C_{m-\overline{m}}^{\overline{m}}}{C_m^{\overline{m}}}$$

where $C_m^{\overline{m}}$ is the number of $\overline{m}$-combinations of m. Given $n = m$, we have

$$C_R^+ - C_A^+ = N_r^+ - N_a^+ \geq PC_m^2 - n(\overline{m} - 1) = C_m^2 \left(1 - \frac{C_{m-\overline{m}}^{\overline{m}}}{C_m^{\overline{m}}}\right) - m(\overline{m} - 1)$$

The term $F = C_m^2 \left(1 - \frac{C_{m-\overline{m}}^{\overline{m}}}{C_m^{\overline{m}}}\right) - m(\overline{m} - 1)$ can be proved to be strictly positive when $\overline{m} \leq m/2$. Hence, $C_R^+ > C_A^+$ in this case. The detailed proof can be found in Appendix L.

For $\overline{m} > m/2$, $P = 1$ and:

$$C_R^+ - C_A^+ = N_r^+ - N_a^+ \geq PC_m^2 - n(\overline{m} - 1) = C_m^2 - m(\overline{m} - 1)$$

Given $n = m$, we have

$$C_R^+ - C_A^+ \geq m^2/2 + m/2 - m\overline{m}$$

Clearly, when $\overline{m} \leq (m+1)/2$, $C_R^+ - C_A^+ \geq 0$ holds. Specifically, when $\overline{m} = (m+1)/2$, $C_R^+ = C_A^+$.

In conclusion, for $n = m$ and $\overline{m} \leq (m+1)/2$, the difference $C_R^+ - C_A^+ \geq 0$ is satisfied.

Based on Lemmas 10.1–10.3, Theorem 10.2 is established.

Theorem 10.2 For any R, given an equal number of rules and agents (*#rules* = *#agents*), RWPM generates at least as many clauses as AWPM when $\overline{m} \leq (m+1)/2$. Specifically, when an MC-net consists solely of positive value rules, both RWPM and AWPM generate the same number of clauses: $C_R = C_A$.

Proof: For any R, Lemmas 10.1 and 10.2 establish that $C_R^+ \leq C_R$ and $C_A^+ \geq C_A$. Additionally, Lemma 10.3 asserts that, given *#rules* = *#agents* and $\overline{m} \leq (m+1)/2$, $C_R^+ \geq C_A^+$ holds. Consequently, we have the inequalities $C_A \leq C_A^+ \leq C_R^+ \leq C_R$. The equality holds when the MC-net contains only positive value rules and $\overline{m} = (m+1)/2$.

By Theorems 10.1 and 10.2, it is clear that given an equal number of rules and agents, when $\overline{m} \leq (m+1)/2$, agent relation-based WPM (AMPM) achieves more compact encoding. Since MaxSAT-solving time typically increases with more variables and clauses, we deduce that under the condition $\overline{m} \leq (m+1)/2$, AWPM outperforms RWPM. According to the method of generating instances in Chapters 8 and 9, $\overline{m}$ is estimated as $\frac{1-\alpha^m}{1-\alpha}$,[1] with m ranging from 30 to 150 and $\alpha = 0.55$. In this model, $\overline{m}$ is approximately 2.2, which is significantly less than $(m+1)/2$. This explains AWPM's higher efficiency in solving the MC-net-based CSG problem, as shown in Chapter 9. However, when $\overline{m}$ exceeds $(m+1)/2$, $C_R > C_A$ may no longer hold.

Example 10.1 Consider four agents a_1, a_2, a_3, a_4 and four positive value rules:

- $r_1 : a_1 \wedge a_2 \wedge a_3 \wedge \neg a_4 \rightarrow 1$,
- $r_2 : a_1 \wedge a_2 \wedge a_4 \wedge \neg a_3 \rightarrow 2$,
- $r_3 : a_1 \wedge a_3 \wedge a_4 \wedge \neg a_2 \rightarrow 3$,
- $r_4 : a_2 \wedge a_3 \wedge a_4 \wedge \neg a_1 \rightarrow 4$.

In this case, we have $n = m = 4$ and $\overline{m} = 4$, which does not satisfy the condition $\overline{m} \leq (m+1)/2$. Each pair of these rules has an "incompatible" relation, resulting in $N_r = C_4^2 = 6$. In contrast, $N_a = \sum_{r_i \in R} (m_i - 1) = 3 + 3 + 3 + 3 = 12$. Therefore, $N_r < N_a$ holds. Additionally, in this example, $T_r = T_a$ and $|R_n| = 0$, leading to $C_R - C_A = N_r - N_a$. Thus, $C_R < C_A$.

1 See Appendix M for the calculation of $\overline{m}$.

10.3 An Interesting Phenomenon: Analysis on a Special Case

Section 10.2 demonstrates that when *#rules* = *#agents* and an MC-net-based CSG problem includes only positive value rules with $\overline{m} = (m+1)/2$, RWPM and AWPM create the same number of variables and clauses. In this case, the performance difference of these two encodings needs further investigation. In this section, we examine this specific scenario and uncover an intriguing phenomenon.

Figure 10.1 compares the performance of RWPM and AWPM, while Table 10.3 shows the number of variables and clauses produced by each encoding. Surprisingly, despite generating roughly the same number of variables and clauses (as shown in Table 10.3), RWPM solves problem instances significantly faster than AWPM (as depicted in Figure 10.1). For instance, with *#rules* = 150, both RWPM and AWPM generate 11 325 Boolean variables and around 1.66×10^6 clauses, yet the solving times differ markedly. Specifically, RWPM solves instances in merely 3.3 seconds, whereas AWPM takes up to 13 seconds to solve the same problem sets.

To explain such tremendous performance differences in this special case, we analyzed these two encodings comprehensively and discovered that the primary reason lies in how transitive laws are encoded, which greatly affects performance. For RWPM and AWPM, the number of clauses for representing transitive laws are $O\left(n^3\right)$ and $O\left(m^3\right)$, respectively. When *#rules* = *#agents*, i.e., $n = m$, the numbers are equivalent. However, the reasoning process by a WPM solver differs considerably. In RWPM, transitive laws are introduced to characterize the reachability of rules via the "compatible on the same coalition" relation in a feasible rule set. This relation is influenced only by pairs of rules that are genuinely compatible within the same coalition, which usually constitutes a smaller subset of all possible rule

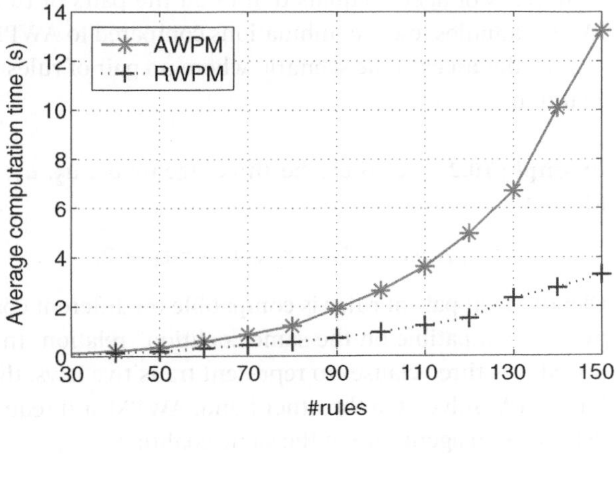

Figure 10.1 Average computation times of AWPM and RWPM given $p = 0.2, q = 0, \alpha = \frac{m-1}{m+1}$ (Source: Xiaojuan Liao et al. (2019)/Oxford University Press).

Table 10.3 Average numbers of variables and clauses in AWPM and RWPM given $p = 0.2, q = 0, \alpha = \frac{m-1}{m+1}$.

#rules	*#variables*		*#clauses*	
	AWPM	RWPM	AWPM	RWPM
30	465	465	12 583.1	12 803.4
40	820	820	30 349.9	30 700.8
50	1275	1275	59 894.5	60 428.3
60	1830	1830	104 250	104 952
70	2485	2485	166 363	167 282
80	3240	3240	249 339	250 399
90	4095	4095	355 946	357 385
100	5050	5050	489 445	491 107
110	6105	6105	652 728	654 689
120	7260	7260	848 775	851 036
130	8515	8515	1 080 684	1 083 219
140	9870	9870	1 351 366	1 354 156
150	11 325	11 325	1 663 805	1 666 923

pairs. Conversely, AWPM deals with $\mathscr{C}_{i,j}$, signifying that agents a_i and a_j are in the same coalition. This relation is influenced by all pairs of agents. Since the number of all pairs of agents equals that of all the pairs of rules when *#agents* = *#rules*, RWPM handles fewer combinations compared to AWPM.

Consider an extreme scenario where no pair of rules is compatible on the same coalition.

Example 10.2 Let there be three agents a_1, a_2, a_3, a_4 and three rules in an MC-net:

$$r_1 : a_1 \wedge \neg a_2 \rightarrow 1, \qquad r_2 : a_2 \wedge \neg a_3 \rightarrow 2, \qquad r_3 : a_3 \wedge \neg a_1 \rightarrow 3$$

Clearly, each pair of rules is compatible on different coalitions; none is reachable by the "compatible on the same coalition" relation. In such cases, while RWPM introduces three clauses to represent transitive laws, these clauses are redundant for a WPM solver. On the other hand, AWPM still requires three clauses to establish that two agents are in the same coalition.

This difference means AWPM ends up with more clauses related to transitive laws than RWPM, leading to a higher computational burden for AWPM.

10.4 RWPM with Refined Transitive Laws (RWPM-RT)

Section 10.3 demonstrated that not all clauses for transitive laws in RWPM are necessary. This section presents a more compact encoding by refining the transitive laws in RWPM.

In RWPM, the compatibility of two rules on the same coalition within a feasible rule set is transitive, represented by the Boolean variable $S_{i,j}$. Furthermore, two rules are considered *reachable* if there exists a chain of $S_{i,j} = 1$ that connects them.

To describe this reachability, we introduce a Boolean variable $H(i,j)$ for each pair of rules r_i and r_j where $i < j$. Here, $H(i,j) = 1$ indicates that r_i and r_j are reachable.

Definition 10.1 ***(RWPM-RT for positive value rules)*** Let R be a set of positive value rules. For each rule $r_i \in R$ (i.e., $r_i : I_i \rightarrow w_i$, $w_i > 0$), we introduce a weighted soft clause (B_i, w_i), where B_i is a Boolean variable. The possible relations on two rules, r_i and r_j $(i < j)$, are encoded into the following hard clauses:

(1) **Compatibility on the same coalition**: If r_i and r_j are compatible on the same coalition, three hard clauses are generated: $\neg B_i \vee \neg B_j \vee H(i,j)$, $\neg H(i,j) \vee B_i$, $\neg H(i,j) \vee B_j$.
(2) **Compatibility on different coalitions**: If r_i and r_j are compatible on different coalitions, then a hard clause is generated: $\neg B_i \vee \neg B_j \vee \neg H(i,j)$.
(3) **Incompatibility**: If r_i and r_j are incompatible, one hard clause is generated: $\neg B_i \vee \neg B_j$.
(4) **Transitive law**: If the relation between r_i and r_j is compatible on different coalitions or independent, then for each rule r_k compatible on the same coalition with r_j, hard clause $\neg H(\{i,k\}) \vee \neg H(\{j,k\}) \vee H(i,j)$ is generated.[2]

Let *Hard* (R) be a set of all the hard clauses mentioned in Definition 10.1. We will prove that a model of *Hard* (R) corresponds with a feasible rule set, and vice versa.

Lemma 10.4 Let R be a set of positive value rules, $R' = \{r'_1, r'_2, \dots, r'_l\} \subseteq R$ be a feasible rule set, and $\mathcal{B}' = \{B'_1, B'_2, \dots, B'_l\}$ be the corresponding Boolean variable set. Then $\mathcal{B}' \cup$ *Hard* (R) is satisfiable.

---

2 $H(\{i,j\}) = H(i,j)$ if $i < j$
$\quad = H(j,i)$ if $j < i$.

Proof: The encodings given in Definitions 10.1-(1)(2)(3) are equivalent to those in Definition 8.5, and the proof follows similarly to Lemma 8.1.

Now we need to verify the correctness of the hard clause introduced by Definition 10.1-(4). For simplicity, assume $i < j < k$.

If $B_i \notin B'$, then $H(i,j) = 0$ and $H(i,k) = 0$ because r_i is not in R'. Hence, the hard clause $\neg H(i,k) \vee \neg H(j,k) \vee H(i,j)$ is satisfied. Similarly, if $B_j \notin B'$ or $B_k \notin B'$, the clause $\neg H(i,k) \vee \neg H(j,k) \vee H(i,j)$ is also satisfied.

Next, consider the case where $B_i \in B'$, $B_j \in B'$, and $B_k \in B'$. If r_i and r_j are compatible on different coalitions, then $H(i,k) = 0$ because $H(j,k) = 1$, since r_i and r_j are not reachable through a "compatible on the same coalition" relation as stated in Theorem 8.1-(2). If r_i and r_j are independent, we need to examine the reachability between r_i and r_k. If r_k is not reachable from r_i, then $H(i,k) = 0$, satisfying $\neg H(i,k) \vee \neg H(j,k) \vee H(i,j)$. If $H(i,k) = 1$, then $H(i,j) = 1$, given $H(j,k) = 1$.

Therefore, the hard clause introduced by Definition 10.1-(4) is satisfied.

Lemma 10.5 Let R be a set of positive value rules and $B' = \{B'_1, B'_2, \dots, B'_l\}$ be a subset of $\mathcal{B}$. If $B' \cup Hard(R)$ is satisfiable, then $\{r'_1, r'_2, \dots, r'_l\}$ is a feasible rule set.

Proof: According to Theorem 8.1, we need to prove that (a) $R' = \{r'_1, r'_2, \dots, r'_l\}$ includes no pair of rules that are incompatible, and (b) if any two rules in R' are compatible on different coalitions, then they are not reachable via the "compatible on the same coalition" relation.

First, we prove that R' includes no pair of rules that are incompatible. Suppose for contradiction that $\exists r'_i, r'_j \in R'$ such that r'_i and r'_j are incompatible. Given that r'_i and r'_j are in R', we have $B'_i = 1$ and $B'_j = 1$. According to Definition 10.1-(3), the clause $\neg B_i \vee \neg B_j$ must be false. This contradicts the assumption that $Hard(R)$ is satisfiable.

Second, assume $\exists r'_i, r'_j \in R'$ such that r'_i and r'_j are compatible on different coalitions and $H(i,j) = 1$. Based on Definition 10.1-(2), either B'_i or B'_j is false. This contradicts the assumption that B' is satisfiable.

Hence, both (a) and (b) are satisfied, confirming that R' is a feasible rule set.

Theorem 10.3 The encoding provided in Definition 10.1 ensures a WPM solver to produce the correct results of the CSG problem that consists of a set of positive value rules.

Proof: According to Lemmas 10.4 and 10.5, a set R' ($\subseteq R$) of rules is feasible if and only if $B' \cup Hard(R)$ is satisfiable, where B' is the set of Boolean variables corresponding to R'. Recall that we have soft clauses (B_i, w_i) for each $B_i \in \mathcal{B}$, and the WPM solver's objective is to find an assignment that satisfies $B' \cup Hard(R)$ and maximizes $\sum_{i=1} w_i B'_i$. Therefore, Theorem 10.3 holds.

Clearly, compared with Definition 8.5, Definition 10.1 generates transitive rules only for case (4), leading to a substantial reduction in the number of clauses required to represent transitive laws.

Example 10.3 Let there be four agents a_1, a_2, a_3, a_4 and four positive value rules:

$$r_1 : a_1 \wedge a_2 \rightarrow 2, \qquad r_2 : a_2 \wedge a_3 \wedge \neg a_4 \rightarrow 2,$$
$$r_3 : a_1 \wedge a_4 \rightarrow 1, \qquad r_4 : a_3 \wedge \neg a_2 \rightarrow 3$$

We introduce four soft clauses $(B_1, 2)$, $(B_2, 2)$, $(B_3, 1)$, and $(B_4, 3)$, respectively. Their rule relations are captured by the following clauses:

- r_1 and r_2 are compatible on the same coalition: $\neg B_1 \vee \neg B_2 \vee H(1,2)$, $\neg H(1,2) \vee B_1$, $\neg H(1,2) \vee B_2$.
- r_1 and r_3 are compatible on the same coalition: $\neg B_1 \vee \neg B_3 \vee H(1,3)$, $\neg H(1,3) \vee B_1$, $\neg H(1,3) \vee B_3$.
- r_1 and r_4 are compatible on different coalitions: $\neg B_1 \vee \neg B_4 \vee \neg H(1,4)$.
- r_2 and r_3 are compatible on different coalitions: $\neg B_2 \vee \neg B_3 \vee \neg H(2,3)$.
- r_2 and r_4 are incompatible: $\neg B_2 \vee \neg B_4$.
- r_3 and r_4 are independent and there is no rule compatible with r_4 on the same coalition: no clause is generated.

The four rules yield only one hard clause for describing the transitive laws, described as follows:

- r_2 and r_3 are compatible on different coalitions, and there is one rule, i.e., r_1, compatible with r_3 on the same coalition: $\neg H(1,2) \vee \neg H(1,3) \vee H(2,3)$.

Altogether, to solve this problem, 8 variables and 14 clauses are generated. The complete file in WPM input format is provided in Appendix N. All the clauses are connected by logical conjunction and input to a WPM solver. The solver then outputs the optimized social welfare and the corresponding truth assignment, which allows for the derivation of the optimal feasible rule set. The assigned truth values of all the Boolean variables are listed in Figure 10.2.

By contrast, regardless of the specific types of rule relations, RWPM generates three clauses for each pair of rules to represent transitive laws, resulting in a total of $4 \times 3 \times 2 \div 2 = 12$ hard clauses. Evidently, the number of clauses generated by RWPM-RT is significantly reduced. For a better comparison, readers may refer to Example 8.9, which provides the complete RWPM encoding for this instance.

$B_1 = 1$	$B_3 = 1$	$H(1,2) = 0$	$H(1,4) = 0$
$B_2 = 0$	$B_4 = 1$	$H(1,3) = 1$	$H(2,3) = 0$

Figure 10.2 Assignment of all Boolean variables in Example 10.3 (Source: Xiaojuan Liao and Miyuki Koshimura/original).

So far, we have discussed RWPM-RT encoding with positive value rules. When an MC-net includes negative value rules, the encoding strategy needs adjustment for two reasons. First, WPM solvers only handle clauses with positive weights [7], necessitating the conversion of negative values to positive. Second, without specific constraints, a solver might disregard negative value rules since selecting them reduces social welfare. This can lead to incorrect results because, in certain situations, satisfying a negative value rule is the prerequisite for satisfying more positive value rules.

To address this, Ueda et al. [3] introduced zero-valued dummy rules for each negative value rule, demonstrating that a negative value rule applies to a coalition if and only if none of its dummy rules apply to any coalition. This constraint is encoded in WPM formulae, with each negative value rule being transformed into a positive value rule [4, 8]. For more details on encoding negative value rules, refer to Section 8.3.3. Example 10.4 illustrates how to apply RWPM-RT to an MC-net-based CSG problem involving both positive and negative value rules.

Example 10.4 Let there be four agents a_1, a_2, a_3, a_4 and three rules:

$$r_1 : a_1 \wedge a_2 \wedge \neg a_4 \rightarrow -1, \quad r_2 : a_1 \wedge a_2 \wedge a_3 \wedge \neg a_4 \rightarrow 4, \quad r_3 : a_2 \wedge a_4 \rightarrow 2$$

We introduce three weighted soft clauses $(\neg B_1, 1)$, $(B_2, 4)$, and $(B_3, 2)$, respectively for r_1, r_2, and r_3. r_1 generates two dummy rules: $r_4 : a_1 \wedge \neg a_2 \rightarrow 0$ and $r_5 : a_1 \wedge a_4 \rightarrow 0$. For r_4 and r_5, Boolean variables B_4 and B_5 are introduced. The constraint imposed on r_1 and its dummy rules is captured by $B_1 \vee B_4 \vee B_5$, and the rule relations are captured by the following clauses:

- r_1 and r_2 are compatible on the same coalition: $\neg B_1 \vee \neg B_2 \vee H(1,2)$, $\neg H(1,2) \vee B_1$, $\neg H(1,2) \vee B_2$.
- r_3 and r_5 are compatible on the same coalitions: $\neg B_3 \vee \neg B_5 \vee H(3,5)$. $\neg H(3,5) \vee B_3$, $\neg H(3,5) \vee B_5$.
- r_4 and r_5 are compatible on the same coalition: $\neg B_4 \vee \neg B_5 \vee H(4,5)$, $\neg H(4,5) \vee B_4$, $\neg H(4,5) \vee B_5$.
- r_3 and r_4 are compatible on different coalitions: $\neg B_3 \vee \neg B_4 \vee \neg H(3,4)$.
- r_1 and r_3 are incompatible: $\neg B_1 \vee \neg B_3$.
- r_1 and r_4 are incompatible: $\neg B_1 \vee \neg B_4$.
- r_1 and r_5 are incompatible: $\neg B_1 \vee \neg B_5$.
- r_2 and r_3 are incompatible: $\neg B_2 \vee \neg B_3$.
- r_2 and r_4 are incompatible: $\neg B_2 \vee \neg B_4$.
- r_2 and r_5 are incompatible: $\neg B_2 \vee \neg B_5$.

$$\begin{array}{lll} B_1 = 1 & B_4 = 0 & H(3,5) = 0 \\ B_2 = 1 & B_5 = 0 & H(4,5) = 0 \\ B_3 = 0 & H(1,2) = 1 & H(3,4) = 0 \end{array}$$

Figure 10.3 Assignment of all Boolean variables in Example 10.4 (Source: Xiaojuan Liao and Miyuki Koshimura/original).

The five rules yield only one hard clause for describing the transitive laws, described as follows:

- r_3 and r_4 are compatible on different coalitions, and there is one rule, i.e., r_5, compatible with r_4 on the same coalition: $\neg H(3,5) \vee \neg H(4,5) \vee H(3,4)$.

Altogether, to solve this problem, 9 variables and 21 clauses are generated. For the complete file in WPM input format, readers may refer to Appendix O.

All the clauses are connected by logical conjunction and input to a WPM solver. The solver then outputs the optimized social welfare and the corresponding truth assignment, which allows for the derivation of the optimal feasible rule set. The assigned truth values of all the Boolean variables are listed in Figure 10.3.

By contrast, regardless of the specific types of rule relations, RWPM generates three clauses for each pair of rules to represent transitive laws, resulting in a total of $5 \times 4 \times 3 \div 2 = 30$ hard clauses. Evidently, the number of clauses generated by RWPM-RT is significantly reduced. For a better comparison, readers may refer to Example 8.12, which provides the complete RWPM encoding for this instance.

10.5 Performance Evaluation

This section conducts a series of comparative experiments to demonstrate the advanced performance of the WPM encoding with refined transitive laws (RWPM-RT). All the encodings in this section were tested on a 2.0 GHz quad-core Intel i7-4510U processor with 8 GB RAM.

10.5.1 Results in a Special Case

In this subsection, RWPM-RT was tested on the same problem instances as in Section 10.3. Results are shown in Figures 10.4 and 10.5. Compared with RWPM and AWPM, RWPM-RT significantly enhances the efficiency of solving the CSG problem. This improvement is mainly due to its more compact encoding, particularly the substantial reduction in the number of clauses (Figure 10.5b). Specifically, when $\#rules = 150$, both AWPM and RWPM generate over 10^6 clauses, while RWPM-RT, with its refined transitive law encoding, generates fewer than 10^5 clauses. This significant reduction in clauses is the primary reason for RWPM-RT's increased efficiency.

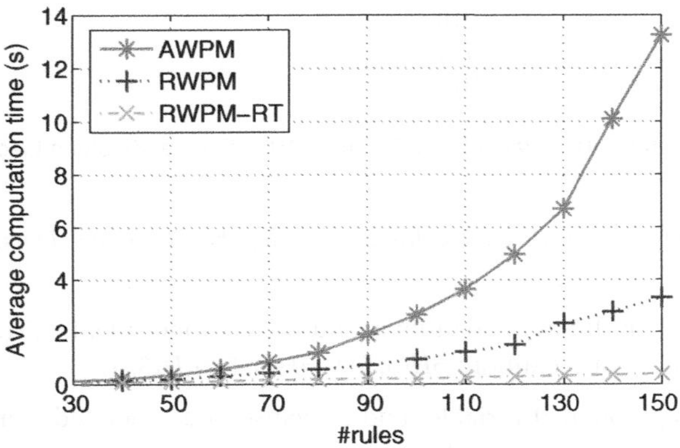

Figure 10.4 Average computation times of AWPM, RWPM and RWPM-RT given $p = 0.2$, $q = 0$, $\alpha = \frac{m-1}{m+1}$ (Source: Xiaojuan Liao et al. (2019)/Oxford University Press).

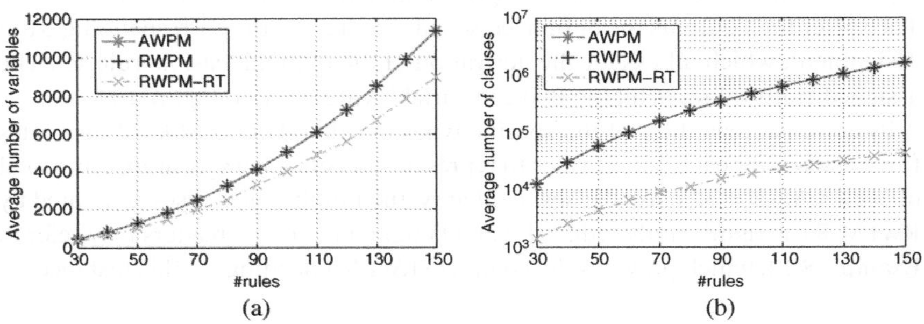

Figure 10.5 Number of variables and clauses in AWPM, RWPM and RWPM-RT given $p = 0.2, q = 0, \alpha = \frac{m-1}{m+1}$. (a) Number of Boolean variables and (b) number of clauses (Source: Xiaojuan Liao et al. (2019)/Oxford University Press).

10.5.2 Results in a General Case

In this subsection, RWPM-RT was tested using the parameter settings consistent with those in Chapters 8 and 9. Specifically, the number of rules was set equal to the number of agents (i.e., *#rules* = *#agents*), with approximately 20% of the rules having negative values and 20% of the literals in a coalition being negative. On average, each condition involved around 2.2 agents.

Figure 10.6 shows the WPM-solving time for three encodings, with arrows indicating the percentage of instances solved within the time limit (omitted if all were solved). Clearly, as seen in Figure 10.6, despite negative value rules in the MC-net, RWPM-RT constantly outperforms RWPM and AWPM as the number of rules (*#rules*) increases.

Additionally, we investigated the number of Boolean variables and clauses in these encodings. As depicted in Figure 10.7a, RWPM-RT generates moderately

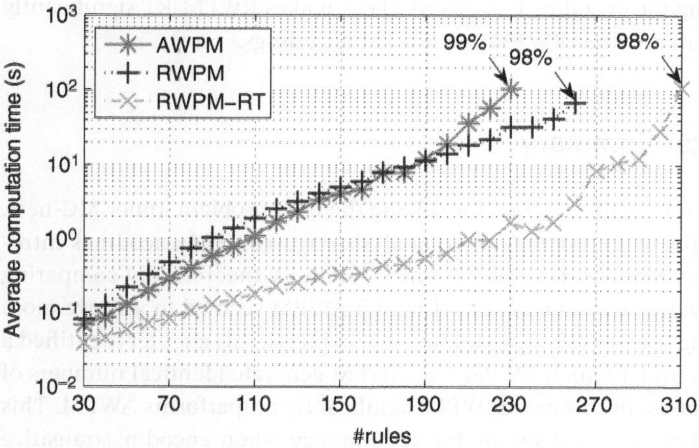

Figure 10.6 Average computation times of AWPM and RWPM-RT given $p = 0.2$, $q = 0.2$, $\alpha = 0.55$ (Source: Xiaojuan Liao et al. (2019)/Oxford University Press).

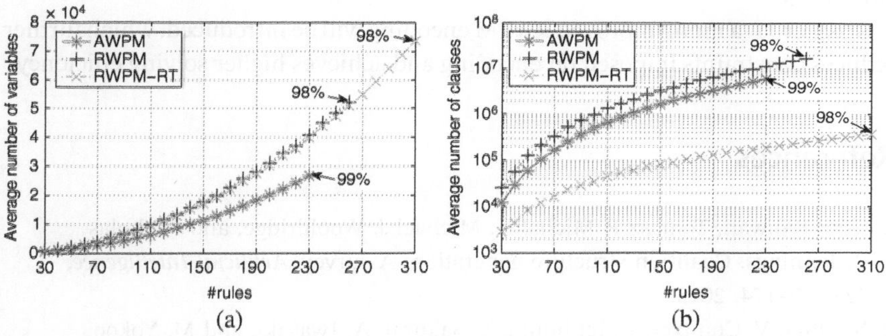

Figure 10.7 Number of variables and clauses in AWPM and RWPM-RT given $p = 0.2$, $q = 0.2$, and $\alpha = 0.55$. (a) Number of Boolean variables and (b) number of clauses (Source: Xiaojuan Liao et al. (2019)/Oxford University Press).

more Boolean variables than AWPM due to the effect of negative value rules. At *#rules* = 230, AWPM and RWPM-RT generate approximately 2.6×10^4 and 4×10^4, respectively, with both orders of magnitude being the same. In contrast, as shown in Figure 10.7b, RWPM-RT generates significantly fewer clauses compared to AWPM and RWPM. At *#rules* = 230, AWPM and RWPM produce 6×10^6 and 10^7 clauses, respectively, while RWPM-RT generates only 2×10^5 clauses. This substantial reduction offsets the impact of the larger number of variables and is the key factor contributing to RWPM-RT's efficiency, explaining its superior performance.

Overall, refining the encoding of transitive laws makes RWPM-RT significantly more scalable than previous RWPM and AWPM encodings.

10.6 Chapter Summary

WPM encodings are highly efficient for solving the CSG problem under MC-nets. This chapter begins with a comparative analysis on two WPM encodings introduced in Chapters 8 and 9, i.e., RWPM and AWPM, by theoretically comparing the numbers of variables and clauses. Although AWPM seemed to achieve more compact encoding than RWPM as exhibited in Chapter 9, this chapter identified a special condition under which RWPM and AWPM generate identical numbers of variables and clauses. In this case, RWPM significantly outperforms AWPM. This performance difference is caused by the redundancy when encoding transitive laws in RWPM. By removing these redundant clauses, we developed an improved RWPM with refined transitive laws and demonstrate that it is more time efficient than the previous encodings since it introduces significantly fewer clauses for transitive laws.

Chapter 11, a more improved RWPM encoding will be introduced, which further reduces constraints imposed on encoding and achieves higher solving efficiency.

References

1 Talal Rahwan, Tomasz P. Michalak, Michael J. Wooldridge, and Nicholas R. Jennings. Coalition structure generation: A survey. *Artificial Intelligence*, 229:139–174, 2015.

2 N. Ohta, V. Conitzer, R. Ichimura, Y. Sakurai, A. Iwasaki, and M. Yokoo. Coalition structure generation utilizing compact characteristic function representation. In *Proceedings of 15th International Conference on Principles and Practice of Constraint Programming (CP '09)*, 2009.

3 S. Ueda, T. Hasegawa, N. Hashimoto, N. Ohta, A. Iwasaki, and M. Yokoo. Handling negative value rules in MC-net-based coalition structure generation. In *Proceedings of 11th International Joint Conference on Autonomous Agents and Multiagent Systems*, 2012.

4 Xiaojuan Liao, Miyuki Koshimura, Hiroshi Fujita, and Ryuzo Hasegawa. Solving the coalition structure generation problem with MaxSAT. In *IEEE 24th International Conference on Tools with Artificial Intelligence, ICTAI 2012*, Athens, Greece, November 7-9, 2012, pages 910–915. IEEE Computer Society, 2012.

5 Xiaojuan Liao, Miyuki Koshimura, Hiroshi Fujita, and Ryuzo Hasegawa. Extending MaxSAT to solve the coalition structure generation problem with

externalities based on agent relations. *IEICE Transactions on Information & Systems*, 97-D(7):1812–1821, 2014.

6 Xiaojuan Liao and Miyuki Koshimura. A comparative analysis and improvement of encodings for coalition structure generation under MC-nets. *Journal of Logic and Computation*, 29(6):913–931, 2019.

7 Chu Min Li and Felip Manyà. MaxSAT, hard and soft constraints. In Armin Biere, Marijn Heule, Hans van Maaren, and Toby Walsh, editors, *Handbook of Satisfiability* - Second Edition, volume 336 of *Frontiers in Artificial Intelligence and Applications*, pages 903–927. IOS Press, 2021.

8 Xiaojuan Liao, Miyuki Koshimura, Hiroshi Fujita, and Ryuzo Hasegawa. MaxSAT encoding for MC-net-based coalition structure generation problem with externalities. *IEICE Transactions on Information & Systems*, 97-D(7):1781–1789, 2014.

11

Improved Rule Relation-Based WPM (I-RWPM)

After reading this chapter, you should be able to:

- Refine the coalition structure generation (CSG) problem based on marginal contribution networks (MC-net) by decomposing it into two subproblems, thereby reducing the problem size and simplifying the overall solution process.
- Design a more compact encoding for rule relations and transitive laws on reduced MC-nets.
- Conduct a comparative analysis of two rule relation-based weighted partial MaxSAT (WPM) encodings.

11.1 Motivation

Before diving into this chapter, we strongly recommend that readers familiarize themselves with the concepts of the rule relation-based weighted partial MaxSAT (RWPM) introduced in Chapter 8 and the RWPM encoding with refined transitive laws (RWPM-RT) introduced in Chapter 10. A solid understanding of the four types of rule relations and their corresponding encoding theories will be particularly beneficial.

The RWPM encoding is far from compact. The redundancy stems from the following two sources:

Firstly, as specified in Definition 8.5, typically, early each pair of rules with any relation other than "independent" requires at least one clause for encoding. However, there are situations where encoding the relation between rules is unnecessary, even when their relation is not "independent." For instance, consider an MC-net with three rules:

$$r_1 : a_1 \wedge a_2 \rightarrow 2, \qquad r_2 : a_2 \wedge \neg a_1 \rightarrow 2, \qquad r_3 : a_3 \wedge \neg a_2 \rightarrow 1$$

Applied Satisfiability: Cryptography, Scheduling, and Coalitional Games,
First Edition. Xiaojuan Liao and Miyuki Koshimura.

Here, $P_1 = \{a1, a2\}$, $N_1 = \emptyset$, $P_2 = \{a_2\}$, $N_2 = \{a_1\}$, $P_3 = \{a_3\}$, $N_3 = \{a_2\}$. Since $P_1 \cap P_2 \neq \emptyset$ and $P_1 \cap N_2 \neq \emptyset$, r_1 and r_2 are incompatible, as defined in Definition 8.4. Similarly, r_3 is compatible on different coalitions with both r_1 and r_2. In this case, the inclusion or exclusion of r_3 in a feasible rule set does not affect the inclusion or exclusion of r_1 and r_2, making the constraints involving r_3 redundant.

Secondly, RWPM generates an excessive number of clauses to encode transitive laws. Specifically, for any three rules, RWPM generates three clauses, leading to a total of $n \cdot (n-1) \cdot (n-2)/2$ clauses for n rules. However, as revealed in Chapter 10, not all these clauses are necessary. For example, consider the following MC-net with three rules:

$$r_1 : a_1 \wedge a_2 \wedge \neg a_3 \rightarrow 2, \qquad r_2 : a_2 \wedge a_3 \wedge \neg a_1 \rightarrow 2,$$
$$r_3 : a_1 \wedge a_3 \wedge \neg a_2 \rightarrow 1$$

Since all rule pairs are incompatible, no transitive clauses are required.

To address encoding redundancy, Liao et al. [1] introduced the improved RWPM (I-RWPM), which offers a more compact encoding. This chapter focuses on I-RWPM and its two main advancements.

First, I-RWPM identifies a subset of rules known as *freelance rules* from an MC-net. The MC-net is then divided into two subsets: one containing only the

Table 11.1 Notations and descriptions used throughout Chapter 11.

Notation	Description
R	A finite set of rules in an MC-net.
r_i	A rule in R.
I_i	The pattern of r_i.
w_i	The value of r_i.
R_f	The set of freelance rules in R.
R_f^+	The set of all positive value rules in R_f.
R_f'	The optimal feasible rule set over R_f.
R_c	The complementary set of R_f over R.
R_c'	The optimal feasible rule set over R_c.
R'	The optimal feasible rule set over R.
B_i	A Boolean variable introduced for r_i.
$H(i,j)$	A Boolean variable introduced by I-RWPM to describe the relation between r_i and r_j. $H(i,j) = 1$ means that r_i and r_j are reachable via the relation of "compatible on the same" coalition in a feasible rule set.

freelance rules, which can be solved without additional encoding, and the other, a reformulated MC-net without these freelance rules. This separation simplifies the problem by reducing the size of the MC-net that requires encoding.

Second, I-RWPM enhances the WPM encoding for the reduced MC-net by minimizing the number of clauses related to rule relations and transitive laws. This improvement further enhances the efficiency of solving the coalition structure generation (CSG) problem. Table 11.1 lists all the notations used in this chapter.

11.2 Identify Freelance Rules in an MC-Net

In an MC-net, Ohta et al. [2] categorized the relationships between pairs of rules into four nonoverlapping and exhaustive cases, as explained in Definition 8.4. I-RWPM utilizes this classification and identifies a special subset of rules within an MC-net, referred to as freelance rules. This section will first define what freelance rules are and then illustrate how these rules aid in simplifying the CSG problem.

Definition 11.1 A rule $r \in R$ is identified as a freelance rule in R if it satisfies the following two conditions:

(a) There is no rule incompatible with r in R.
(b) There is no rule compatible on the same coalition with r in R.

In simpler terms, a rule $r \in R$ qualifies as a freelance rule in R if r meets the following condition: for every $r' \in R \backslash \{r\}$, r' is either independent of r or compatible with r on different coalitions.

Next, we introduce two fundamental facts about freelance rules that will assist in solving the CSG problem.

Lemma 11.1 If a rule r is a freelance rule in R, any feasible rule set R' ($\subset R$) that does not include r can be expanded to include r without affecting the feasibility of R'.

Proof: According to Theorem 8.1, no rules in R' have the relation of "incompatible." Since r is a freelance rule in R, adding it to R' to form $R' \cup \{r\}$ does not introduce any incompatibility. Additionally, the reachability of rules in R' via "compatible on the same coalition" edges remains unchanged when r is included. Consequently, if two rules in $R' \cup \{r\}$ are "compatible on different coalitions," they are not connected via "compatible on the same coalition" edges in $R' \cup \{r\}$. Thus, $R' \cup \{r\}$ satisfies both conditions of Theorem 8.1, confirming that $R' \cup \{r\}$ is a feasible rule set.

Lemma 11.2 Given R_f as the set of all freelance rules in R and R_f^+ as the subset of R_f containing all positive value rules, formally defined as $R_f^+ = \{r_i | r_i \in R_f, w_i > 0\}$, it follows that R_f^+ is the optimal feasible rule set within R_f.

Proof: First, we prove that R_f^+ is a feasible rule set. Consider a positive value rule r in R_f, i.e., $r \in R_f^+$. Clearly, $\{r\}$ is feasible since we can always create a coalition structure $CS = \{\{r\}, A \backslash \{r\}\}$. By applying Lemma 11.1, this feasible rule set $\{r\}$ can be extended by adding any freelance rules, which implies that R_f^+ is feasible.

Next, we establish that the value derived from R_f^+ is maximized, i.e., $\sum_{r_i \in R_f^+} w_i$ is the maximum. This follows directly because R_f^+ includes all positive value rules in R_f, ensuring that for any subset $R_x \subseteq R_f$, $\sum_{r_i \in R_f^+} w_i \geq \sum_{r_i \in R_x} w_i$.

Lemma 11.1 shows that freelance rules can be added into or removed from an arbitrary feasible rule set without changing the feasibility of the rule set. Additionally, Lemma 11.2 points out that determining the optimal feasible rule set among freelance rules is straightforward, i.e., it is simply the collection of all positive value rules in R_f. Now, we will illustrate how freelance rules can simplify the process of solving the CSG problem.

Theorem 11.1 Let R be a set of rules, R_f the set of all freelance rules in R, and R_c the set that complements R_f within R, i.e., $R_c = R \backslash R_f$. Suppose R' represents the optimal feasible rule set over R. If $R'_c = R' \cap R_c$ and $R'_f = R' \cap R_f$, then R'_c is the optimal feasible rule set over R_c, and $R'_f = R_f^+$.

Proof: Since $R'_c = R' \cap R_c$, it follows that R'_c is a subset of both R' and R_c. Given that R' is the optimal feasible rule set over R, it is evident that R'_c must be a feasible rule set over R_c. Similarly, R'_f can be shown to be a feasible rule set over R_f.

Next, we will demonstrate that R'_c and R'_f are optimal over R_c and R_f, respectively. Assume, for the sake of contradiction, that R'_c is not optimal over R. Then there must exist another feasible rule set R_c^* over R_c such that $R_c^* \neq R'_c$ and $\sum_{r_i \in R'_c} w_i < \sum_{r_i \in R_c^*} w_i$. By Lemma 11.1, any feasible rule set can be extended to include freelance rules. Therefore, adding R'_f to R_c^* forms a new feasible rule set $R^* = R_c^* \cup R'_f$ over R. Clearly, $\sum_{r_i \in R'} w_i < \sum_{r_i \in R^*} w_i$, which contradicts the assumption that R' is optimal. Hence, R'_c must be optimal over R_c. Similarly, we can prove that R'_f is optimal over R_f. According to Lemma 11.2, the optimal feasible rule set over R_f is R_f^+, so $R'_f = R_f^+$.

Based on Theorem 11.1, solving the MC-net-based CSG problem can be approached by addressing two simplified subproblems. One is to identify the positive value rules among the freelance rules, and the other one is to determine the optimal feasible rule set from the remaining rules. To be specific, suppose R is an MC-net, the process of determining the optimal feasible rule set over R is as follows:

(1) Construct R_f as the set of all freelance rules in R, and R_c as the complement of R_f in R, i.e., $R \backslash R_f$.
(2) Determine R_f^+, which includes all positive value rules from R_f, i.e., $R_f^+ = \{r_i | r_i \in R_f, w_i > 0\}$.
(3) Compute the optimal feasible rule set R_c' over R_c.
(4) The optimal feasible rule set R' for R is the union of R_f^+ and R_c', i.e., $R' = R_f^+ \cup R_c'$.

Notably, steps (1), (2), and (4) are straightforward and can be performed without any additional encoding. As a result, solving the optimal feasible rule set over R simplifies to solving it over R_c, which is likely smaller since $R_c \subseteq R$. The encoding strategy for step (3) will be detailed in Section 11.3.

Example 11.1 Let us consider a rule set R with five agent a_1, a_2, a_3, a_4, a_5 and five rules:

$$r_1 : a_1 \wedge a_2 \to 2, \quad r_2 : a_1 \wedge a_4 \wedge \neg a_2 \to 2, \quad r_3 : a_2 \wedge a_4 \to 1,$$

$$r_4 : a_3 \wedge \neg a_2 \to 3, \quad r_5 : a_5 \wedge \neg a_4 \to -1$$

The relationships between these rules are as follows:

- r_1 and r_2 are incompatible.
- r_1 and r_3 are compatible on the same coalition.
- r_1 and r_4 are compatible on different coalitions.
- r_1 and r_5 are independent.
- r_2 and r_3 are incompatible.
- r_2 and r_4 are independent.
- r_2 and r_5 are compatible on different coalitions.
- r_3 and r_4 are compatible on different coalitions.
- r_3 and r_5 are compatible on different coalitions.
- r_4 and r_5 are independent.

Figure 11.1a shows a graphical representation of this example. Each rule is represented by a vertex, and edges between vertices indicate the type of rule relation, with "independent" relations not shown.

As can be seen in Figure 11.1a, neither r_4 nor r_5 is connected by "compatible on the same coalition" or "incompatible" edges. According to Definition 11.1, r_4 and r_5 are freelance rules. Therefore, $R_f = \{r_4, r_5\}$, $R_c = R \backslash R_f = \{r_1, r_2, r_3\}$, and $R_f^+ = \{r_4\}$. By Theorem 11.1, the optimal feasible rule set R' is given by $R' = R_c' \cup R_f^+ = R_c' \cup \{r_4\}$.

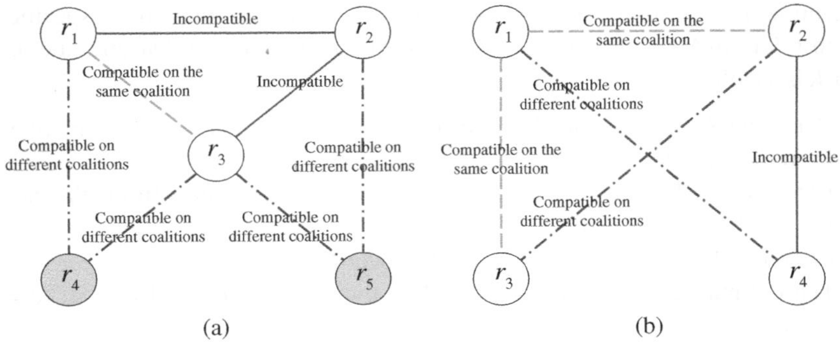

Figure 11.1 Graphical representation of examples (a) Example 11.1 and (b) Example 11.2 (Source: (a) Xiaojuan Liao et al. (2019)/Springer Nature; (b) Xiaojuan Liao et al./original).

The problem thus reduces to finding the optimal feasible rule set for $R_c = \{r_1, r_2, r_3\}$. Compared to the original problem of finding the optimal feasible rule set for $R = \{r_1, r_2, r_3, r_4, r_5\}$, the problem has been scaled down.

Example 11.2 Let us consider a rule set R with four agents a_1, a_2, a_3, a_4 and four rules:

$$r_1 : a_1 \wedge a_2 \rightarrow 2, \qquad r_2 : a_2 \wedge a_3 \wedge \neg a_4 \rightarrow 2,$$
$$r_3 : a_1 \wedge a_4 \rightarrow 1, \qquad r_4 : a_3 \wedge \neg a_2 \rightarrow 3$$

The rule relations over R are identified in Figure 11.1b. In this graph, each rule is connected by either a "compatible on the same coalition" or an "incompatible" edge, indicating that none of these rules in R qualify as freelance rules. Therefore, the processing steps introduced in this section are not applicable in this case.

11.3 Improved Weighted Partial MaxSAT Encoding on Refined MC-Nets

As revealed in Chapter 10, the RWPM encoding (introduced in Chapter 8) generates a large number of clauses to represent transitive laws, many of which are redundant. In Chapter 10, RWPM-RT was presented to reduce these clauses, leading to improved solving efficiency. In this section, we design a more compact encoding that effectively decreases the number of clauses for both rule relations and transitive laws.

The main focus here is on the improved rule relation-based WPM (I-RWPM) encoding, which is applied to the refined MC-net that excludes all freelance rules, corresponding to step (3) in Section 11.2.

11.3.1 I-RWPM Encoding Theory

Consider a set $R = \{r_1, \dots, r_n\}$ consisting of positive value rules, with no freelance rules. For each rule r_i, we define a Boolean variable B_i for $i = 1, \dots, n$, and let $\mathcal{B}$ represent the collection of these variables, i.e., $\mathcal{B} = \{B_1, \dots, B_n\}$. Essentially, $B_i = 1$ signifies that r_i belongs to a feasible rule set.

Among the different types of rule relationships, the "compatible on the same coalition" relation is transitive. This implies that if r_i and r_j are compatible on the same coalition, and similarly r_j and r_k are compatible on the same coalition, then r_i and r_k are linked through this relation. To characterize this transitivity, as done in Chapter 10, we introduce a Boolean variable $H(i,j)$ for each pair of rules r_i and r_j with $i < j$. Here, $H(i,j) = 1$ indicates that r_i and r_j are *reachable* through the "compatible on the same coalition" relation in a feasible rule set.

Definition 11.2 ***(WPM for reduced MC-net excluding freelance rules)*** Let R be a set of positive value rules *excluding freelance rules.* For every positive value rule $r_i \in R$ (i.e., $r_i : I_i \to w_i,\ w_i > 0$), a weighted soft clause (B_i, w_i) is added, where B_i is a Boolean variable.

The potential relations between two rules r_i and r_j $(i < j)$ are encoded into the following hard clauses:

(1) **Compatibility on the same coalition:** If r_i and r_j are compatible on the same coalition, three hard clauses are generated: $\neg B_i \vee \neg B_j \vee H(i,j)$, $\neg H(i,j) \vee B_i$, $\neg H(i,j) \vee B_j$.
(2) **Incompatibility:** If r_i and r_j are incompatible, a hard clause is introduced: $\neg B_i \vee \neg B_j$.
(3) **Compatibility on different coalitions:** If r_i and r_j are compatible on different coalitions, and have rules compatible on the same coalition with themselves, a hard clause is introduced: $\neg B_i \vee \neg B_j \vee \neg H(i,j)$.
(4) **Transitive law:** If r_i and r_j are either compatible on different coalitions or independent, and have rules compatible on the same coalition with themselves, then for each rule r_k that is compatible on the same coalition with r_j, a hard clause is added: $\neg H(\{i,k\}) \vee \neg H(\{j,k\}) \vee H(i,j)$[1]
(5) Otherwise,[2] no clause is introduced.

Let *Hard* (R) be the set of all hard clauses specified in Definition 11.2. We will now prove that a model of *Hard* (R) corresponds to a feasible rule set, and vice versa.

---

1 $H(\{i,j\}) = H(i,j)$ if $i < j$
$\quad = H(j,i)$ if $j < i$.

2 If r_i and r_j are compatible on different coalitions or independent, and there exists no rule compatible on the same coalition with r_i or r_j.

Lemma 11.3 Consider a set R of positive value rules and a feasible rule set $R' = \{r'_1, r'_2, \dots, r'_l\} \subseteq R$ with its corresponding Boolean variable set $B' = \{B'_1, B'_2, \dots, B'_l\}$. Then, $B' \cup Hard(R)$ is satisfiable.

Proof: First, consider the hard clauses introduced by Definition 11.2-(1). If both $B_i \in B'$ and $B_j \in B'$, then $H(i,j) = 1$ because r_i and r_j belong to R' and are compatible on the same coalition. If $B_i \notin B'$, we set $H(i,j) = 0$ since $r_i \notin R'$. Similarly, if $B_j \notin B'$, $H(i,j) = 0$ as well. Thus, the hard clauses from Definition 11.2-(1) are satisfied.

Next, consider the hard clause from Definition 11.2-(2). Suppose both $B_i \in B'$ and $B_j \in B'$. According to Theorem 8.1-(1), either $r_i \notin R'$ or $r_j \notin R'$, which contradicts the assumption. Hence, $\neg B_i \vee \neg B_j$ holds.

For the hard clause introduced by Definition 11.2-(3), if $B_i \notin B'$ or $B_j \notin B'$, the clause $\neg B_i \vee \neg B_j \vee \neg H(i,j)$ is naturally satisfied. Therefore, we only need to consider the case where both B_i and B_j are in B'. In this scenario, $H(i,j) = 0$ because r_i and r_j are not reachable according to Theorem 8.1-(2).

Lastly, consider the hard clause from Definition 11.2-(4). For simplicity, assume $i < j < k$. If $B_i \notin B'$, then $H(i,j) = 0$ and $H(i,k) = 0$ since $r_i \notin R'$. Consequently, the hard clause $\neg H(i,k) \vee \neg H(j,k) \vee H(i,j)$ is satisfied. Similarly, if $B_j \notin B'$ or $B_k \notin B'$, the clause is also satisfied. Now, consider the case where $B_i \in B'$, $B_j \in B'$, and $B_k \in B'$. If r_i and r_j are compatible on different coalitions, we must set $H(i,k) = 0$ given $H(j,k) = 1$ since r_i and r_j are not reachable via the "compatible on the same coalition" relation, as per Theorem 8.1-(2). If r_i and r_j are independent, the reachability between r_i and r_k must be considered. If r_k is not reachable from r_i, then $H(i,k) = 0$, satisfying the clause $\neg H(i,k) \vee \neg H(j,k) \vee H(i,j)$. If $H(i,k) = 1$, we set $H(i,j) = 1$ given that $H(j,k) = 1$. Therefore, the hard clause from Definition 11.2-(4) is satisfied.

Lemma 11.4 Consider a set R of positive value rules and $B' = \{B'_1, B'_2, \dots, B'_l\}$ be a subset of B. If $B' \cup Hard(R)$ is satisfiable, then the corresponding rule set $\{r'_1, r'_2, \dots, r'_l\}$ is a feasible rule set.

Proof: The proof follows by reversing the logic used in Lemma 11.3. Since $B' \cup Hard(R)$ is satisfiable, the rule set $R' = \{r'_1, r'_2, \dots, r'_l\}$ does not contain any pairs of rules that are connected by the "incompatible" relation, as ensured by Definition 11.2-(2). Furthermore, if any two rules in R' are compatible on different coalitions, they are not connected via the "compatible on the same coalition" relation, in line with Definition 11.2-(3). Thus, R' is confirmed to be a feasible rule set.

Theorem 11.2 The encoding described in Definition 11.2 ensures a WPM solver to accurately solve the CSG problem for a set of positive value rules.

Proof: According to Lemma 11.3 and 11.4, a rule set R' ($\subseteq R$) is feasible if and only if the conjunction $\mathcal{B}' \cup Hard(R)$ is satisfiable, where $\mathcal{B}'$ is the set of Boolean variables corresponding to R'. Given that the WPM solver seeks to find an assignment that satisfies $\mathcal{B}' \cup Hard(R)$ while maximizing $\sum_{i=1} w_i B'_i$, Theorem 11.2 holds.

The I-RWPM encoding also adapts to handling negative value rules by introducing dummy rules, as stated in Section 8.3.3 of Chapter 8. Once the negative value rules are encoded, the process follows the same method used for positive value rules, as outlined in Definition 11.2, by identifying relations between each pair of rules and generating corresponding hard clauses. In RWPM and RWPM-RT, dummy rules are created for every negative value rule, regardless of whether the rule is a freelance rule. In contrast, I-RWPM constructs dummy rules only based on negative value rules in the reduced MC-net, which excludes all freelance rules. Specifically, let R_d and R_{dc} be the set of dummy rules generated by negative value rules in an MC-net R and a reduced MC-net R_c, respectively. Rule relations in RWPM and RWPM-RT are encoded over $R \cup R_d$, while in I-RWPM, they are encoded over $R_c \cup R_{dc}$. Clearly, $R_c \subseteq R$ and $R_{dc} \subseteq R_d$. Consequently, the constraints on rule relations specified in I-RWPM are no more than those in RWPM and RWPM-RT.

The I-RWPM encoding can also be tailored to accommodate the concise representation of partition function games (PFGs) [3], where the value of a coalition is influenced by the formation of other coalitions. For details of the generation of auxiliary rules, see Section 8.3.2 in Chapter 8. In PFGs, the constraints on rule relations encoded in Definition 11.2 are extended to auxiliary rules, with no additional logic changes necessary.

11.3.2 Interpretation of I-RWPM

This section begins by illustrating Definition 11.2 using a graphical representation of an MC-net, followed by a detailed comparative analysis of I-RWPM and RWPM to enhance the readers' understanding of encoding theory.

Consider an MC-net represented by a graph G, where each vertex represents a rule, and an edge exists between r_i and r_j if and only if they are compatible on the same coalition in a feasible rule set. In this context, $H(i,j) = 1$ signifies that r_j is directly or indirectly reachable from r_i in G. The four cases in Definition 11.2 can be interpreted as follows:

- **Case (1)**: If r_i and r_j are compatible on the same coalition, an edge exists between them in G if and only if both r_i and r_j are in a feasible rule set.
- **Case (2)**: If r_i and r_j are incompatible, no feasible rule set contains both r_i and r_j. Consequently, $\neg B_i \vee \neg B_j$ holds true.

- **Case (3)**: If r_i and r_j are compatible on different coalitions, r_j cannot be reachable from r_i in G.
- **Case (4)**: If r_k is reachable from r_i and there is an edge between r_k and r_j in G, then r_j is indirectly reachable from r_i.

In case (4), we focus solely on r_k that is compatible with r_j on the same coalition, without considering another rule (say r_x) that is compatible with r_i. This reflects an inherent asymmetry in the reachability analysis. When assessing the reachability between r_i and r_j, we initiate the process from r_i and aim to determine if we can reach r_j. If we cannot establish this connection, we conclude that r_i and r_j are not reachable from each other. Importantly, we never reverse this process by starting from r_j to r_i. The asymmetry is justified because, during the path evaluation from r_i to r_j, the final step involves moving from r_k, which is compatible with r_j on the same coalition, to r_j. This reflects the intuitive meaning of case (4). Hence, there is no need to consider a rule r_x that is compatible with r_i on the same coalition, as such a rule would have already been addressed in the preceding steps.

Definition 11.2 in I-RWPM is derived from Definition 8.5 in RWPM, making these two encodings directly comparable. For a detailed discussion on the comparative analysis of RWPM and I-RWPM, please refer to Appendix P.

11.3.3 Pedagogical Examples

This subsection presents two examples to demonstrate the conciseness of the improved rule relation-based WPM (I-RWPM) encoding. Example 11.3 is a standard instance previously discussed in Chapters 8, 9, and 10, while Example 11.4 introduces a new case that highlights the efficiency of the I-RWPM encoding.

Example 11.3 For the MC-net-based CSG problem in Example 11.2, I-RWPM introduces four soft clauses $(B_1, 2)$, $(B_2, 2)$, $(B_3, 1)$, and $(B_4, 3)$ for r_1, r_2, r_3, and r_4, respectively. Their rule relations are captured by the following clauses:

- r_1 and r_2 are compatible on the same coalition:
  $\neg B_1 \vee \neg B_2 \vee H(1,2)$, $\neg H(1,2) \vee B_1$, $\neg H(1,2) \vee B_2$.
- r_1 and r_3 are compatible on the same coalition:
  $\neg B_1 \vee \neg B_3 \vee H(1,3)$, $\neg H(1,3) \vee B_1$, $\neg H(1,3) \vee B_3$.
- r_2 and r_4 are incompatible: $\neg B_2 \vee \neg B_4$.
- r_2 and r_3 are compatible on different coalitions. Additionally, both r_2 and r_3 are compatible with a rule (i.e., r_1) on the same coalition: $\neg B_2 \vee \neg B_3 \vee \neg H(2,3)$.
- r_1 and r_4 are compatible on different coalitions, and there is no rule compatible with r_4 on the same coalition: no clause is generated.
- r_3 and r_4 are independent and there is no rule compatible with r_4 on the same coalition: no clause is generated.

The four rules yield only one hard clause for describing the transitive laws, described as follows:

- r_2 and r_3 are compatible on different coalitions, and there is one rule, i.e., r_1, compatible with r_2 and r_3 on the same coalition: $\neg H(1,2) \vee \neg H(1,3) \vee H(2,3)$.

Altogether, to solve this problem, 7 variables and 13 clauses are generated. The complete file in WPM input format is provided in Appendix Q.

In contrast, RWPM generates 10 variables and 25 clauses to solve this problem. The primary redundancy comes from the transitive laws. RWPM creates three clauses for each pair of rules to represent these laws, resulting in a total of $4 \times 3 \times 2 \div 2 = 12$ hard clauses.

Comparing I-RWPM with RWPM-RT, we find that RWPM-RT generates 8 variables and 14 clauses. The compatibility of r_1 and r_4 on different coalitions leads RWPM-RT (and also RWPM) to introduce a Boolean variable $H(1,4)$ and the clause $\neg B_1 \vee \neg B_4 \vee \neg H(1,4)$. Conversely, I-RWPM considers them redundant because no rule is compatible with r_4 on the same coalition, making r_4 unreachable from any rule.

For a more detailed comparison, readers may refer to Example 8.9 in Chapter 8 and Example 10.3 in Chapter 10, which provide the complete RWPM and RWPM-RT encodings for this instance.

Example 11.4 Let us consider a rule set R with four agents a_1, a_2, a_3, a_4 and four rules:

$$r_1 : a_1 \wedge a_2 \wedge \neg a_3 \rightarrow 3, \qquad r_2 : a_1 \wedge a_3 \wedge \neg a_2 \rightarrow 2,$$
$$r_3 : a_4 \wedge \neg a_2 \rightarrow 1, \qquad r_4 : a_3 \wedge a_4 \rightarrow 2$$

The graphical representation of this example is depicted in Figure 11.2. I-RWPM introduces four soft clauses $(B_1, 3)$, $(B_2, 2)$, $(B_3, 1)$, and $(B_4, 2)$ for r_1, r_2, r_3, and r_4, respectively. Their rule relations are captured by the following clauses:

- r_2 and r_4 are compatible on the same coalition: $\neg B_2 \vee \neg B_4 \vee H(2,4)$, $\neg H(2,4) \vee B_2$, $\neg H(2,4) \vee B_4$.
- r_3 and r_4 are compatible on the same coalition: $\neg B_3 \vee \neg B_4 \vee H(3,4)$, $\neg H(3,4) \vee B_3$, $\neg H(3,4) \vee B_4$.
- r_1 and r_2 are incompatible: $\neg B_1 \vee \neg B_2$.
- r_1 and r_3 are compatible on different coalitions, and there is no rule compatible with r_1 on the same coalition: no clause is generated.
- r_1 and r_4 are compatible on different coalitions, and there is no rule compatible with r_1 on the same coalition: no clause is generated.

The four rules yield only one hard clause for describing the transitive laws, described as follows:

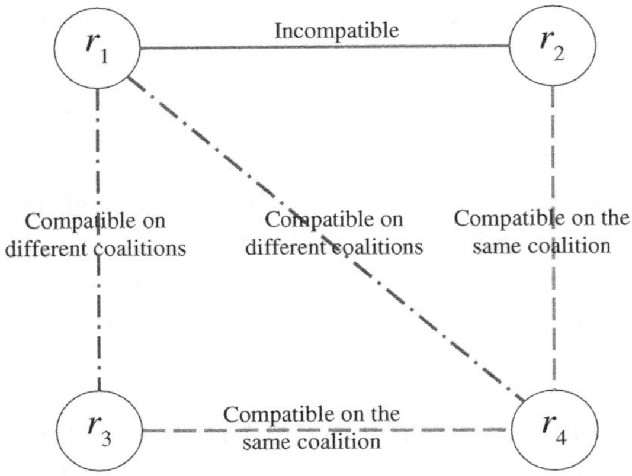

Figure 11.2 Graphical representation of Example 11.4 (Source: Xiaojuan Liao and Miyuki Koshimura/original).

- r_2 and r_3 are independent, and there is one rule, i.e., r_4, compatible with r_2 and r_3 on the same coalition: $\neg H(2,4) \vee \neg H(3,4) \vee H(2,3)$.

Altogether, to solve this problem, 7 variables and 12 clauses are generated by I-RWPM.

In comparison, RWPM-RT generates 10 variables and 17 clauses to address this problem. Specifically, RWPM-RT encodes the rules and the relations of "compatible on the same coalition" and "incompatible" in the same way as I-RWPM. However, it employs different strategies for encoding the "compatible on different coalitions" relation, as illustrated below:

- r_1 and r_3 are compatible on different coalitions: $\neg B_1 \vee \neg B_3 \vee \neg H(1,3)$.
- r_1 and r_4 are compatible on different coalitions: $\neg B_1 \vee \neg B_4 \vee \neg H(1,4)$.

To encode transitive laws, RWPM-RT generates the following clauses:

- r_1 and r_3 are compatible on different coalitions, and there is a rule r_4 compatible with r_3 on the same coalition: $\neg H(1,4) \vee \neg H(3,4) \vee H(1,3)$.
- r_1 and r_4 are compatible on different coalitions, and there are two rules, r_2 and r_3, compatible with r_4 on the same coalition: $\neg H(1,2) \vee \neg H(2,4) \vee H(1,4)$, $\neg H(1,3) \vee \neg H(3,4) \vee H(1,4)$.
- r_2 and r_3 are independent, and there is a rule r_4 compatible with r_3 on the same coalition: $\neg H(2,4) \vee \neg H(3,4) \vee H(2,3)$.

The complete file generated by I-RWPM and RWPM-RT for Example 11.4 in WPM Input Format is provided in Appendix R.

These two examples demonstrate that I-RWPM generates fewer variables and clauses than RWPM and RWPM-RT, even in the absence of freelance rules in an MC-net. When an MC-net includes freelance rules, the advantage of I-RWPM becomes even more significant, as it introduces no clauses for these rules.

11.4 Performance Evaluation

This section compares the performance of the improved rule relation-based WPM (I-RWPM) with three existing WPM encodings: the rule relation-based WPM (RWPM) (in Chapter 8), the agent relation-based WPM (AWPM) (in Chapter 9), and the rule relation-based WPM with refined transitive laws (RWPM-RT) (in Chapter 10). The comparison is conducted across three scenarios:

- A general case that was tested in Chapters 8, 9, and 10.
- Special cases with varying percentage of freelance rules, which ranges from 5% to 25%.
- A special case with a low level of freelance rules, i.e., below 0.05%.

In all experiments, the number of rules is set equal to the number of agents (i.e., *#rules* = *#agents*). All the tests were conducted on a 3.7 GHz quad-core Intel i3-4170 processor with 4GB of RAM.

11.4.1 Results in a General Case

This subsection follows the general parameter settings from Chapters 8, 9, and 10. In this configuration, approximately 20% of the rules have negative values, and 20% of the literals in a coalition are negative. Additionally, each condition contains around 2.2 agents on average.

Figure 11.3 depicts the average computation time for solving the generated problem instances, demonstrating that I-RWPM outperforms the other three encodings across all *#rules* settings.

Figure 11.4 illustrates the numbers of Boolean variables and clauses in the four encodings. As shown in Figure 11.4a, RWPM-RT and I-RWPM are intermediate

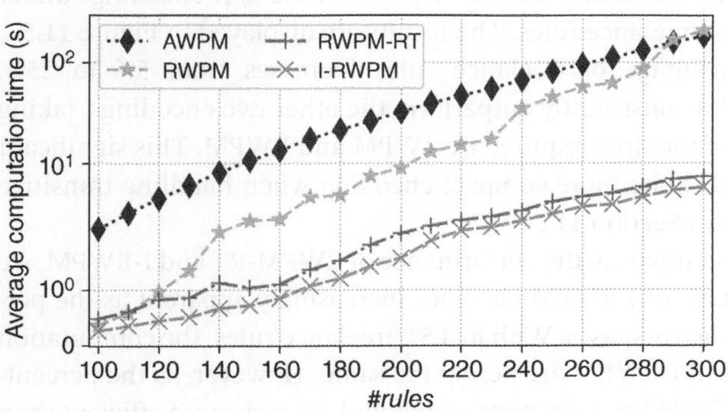

Figure 11.3 Average computation time for AWPM, RWPM, RWPM-RT, and I-RWPM (Source: Xiaojuan Liao and Miyuki Koshimura/original).

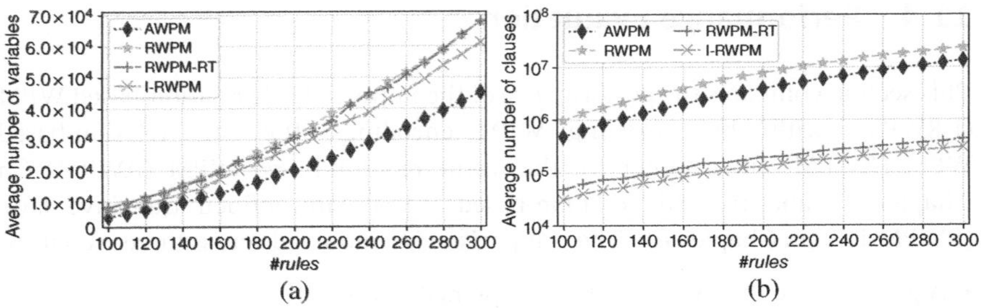

Figure 11.4 Number of variables and clauses in AWPM, RWPM, RWPM-RT, and I-RWPM (a) Number of Boolean variables and (b) Number of clauses (Source: Xiaojuan Liao and Miyuki Koshimura/original).

in the number of Boolean variables. Although these encodings produce different numbers of variables at any fixed *#rules*, they all share the same order of magnitude. In contrast, as shown in Figure 11.4b, the number of clauses generated by RWPM-RT and I-RWPM is significantly smaller than those produced by AWPM and RWPM. This substantial reduction in clauses outweighs the adverse impact of the slightly larger number of variables, making RWPM-RT and I-RWPM significantly more efficient. Furthermore, when comparing RWPM-RT and I-RWPM, I-RWPM generates fewer variables and clauses than RWPM-RT. Given that fewer variables and clauses generally result in faster execution time, this reduction partially accounts for the superior performance of I-RWPM.

11.4.2 Results with Varied Number of Freelance Rules

In this subsection, we constructed problem instances with 5%, 10%, 15%, 20%, and 25% freelance rules to evaluate the performance of the four encodings under different percentages of freelance rules. The results are displayed in Figure 11.5.

Clearly, as the percentage of freelance rules increases from 5% to 25%, RWPM-RT and I-RWPM consistently outperform the other two encodings, taking only about one-tenth of the time required by AWPM and RWPM. This significant advantage is attributed to the more compact encoding when handling transitive relations (as described in Section 11.3).

Furthermore, when comparing the performance of RWPM-RT and I-RWPM, we notice that the advantage of I-RWPM becomes increasingly apparent as the percentage of freelance rules increases. With just 5% freelance rules, the computation times for RWPM-RT and I-RWPM are nearly the same. However, as the percentage climbs to 25%, I-RWPM becomes approximately 1.5 times more efficient than RWPM-RT. This highlights the importance of identifying freelance rules to boost the overall efficiency (as described in Section 11.2).

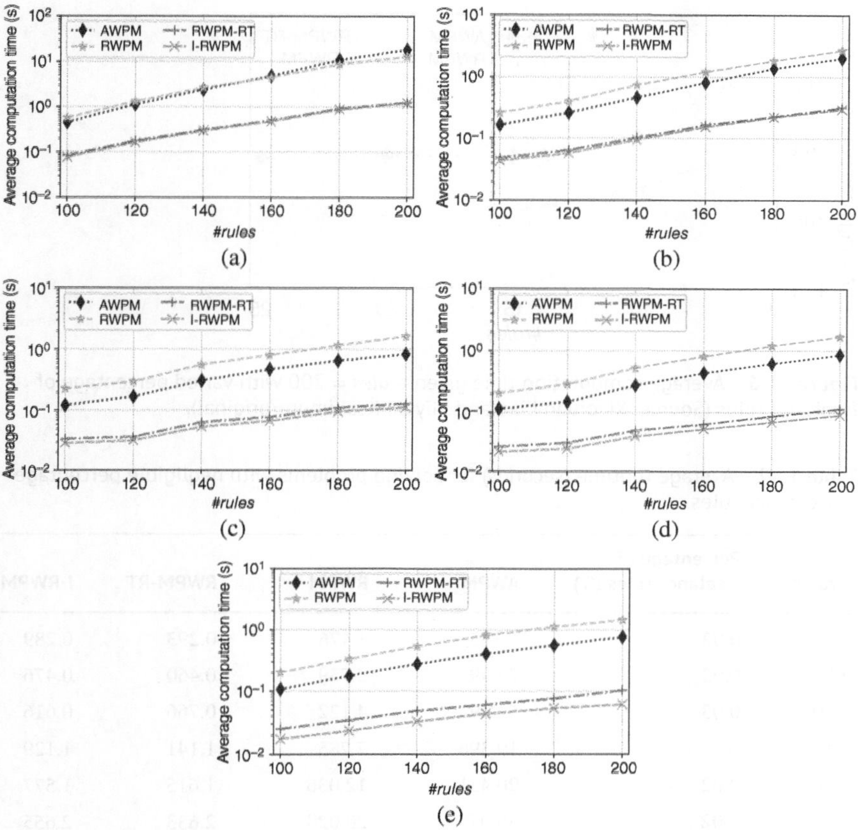

Figure 11.5 Average computation time with specific number of freelance rules (a) 5% freelance rules, (b) 10% freelance rules, (c) 15% freelance rules, (d) 20% freelance rules, and (e) 25% freelance rules (Source: Xiaojuan Liao and Miyuki Koshimura/original).

Another notable observation is that as the proportion of freelance rules rises, the problem becomes less complex. For example, with 25% freelance rules, all encodings show a significant improvement in efficiency, performing 10 times faster than with 5% freelance rules, as depicted in Figure 11.6. This suggests that the decreasing trend seen in Figure 11.6 would likely continue with an even higher percentage of freelance rules.

11.4.3 Results with Few Freelance Rules

This subsection examines a special case where the percentage of freelance rules in an MC-net is extremely low, specifically no more than 0.05%. Table 11.2

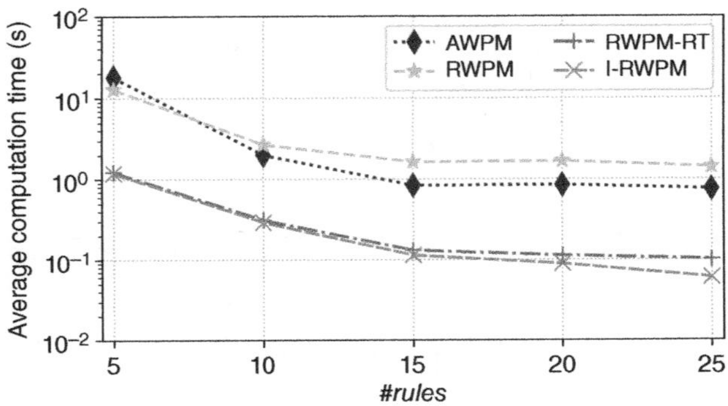

Figure 11.6 Average computation time given *#rules* = 200 with varied percentage of freelance rules (Source: Xiaojuan Liao and Miyuki Koshimura/original).

Table 11.2 Average runtime (seconds) for solving problems with negligible percentage of freelance rules.

#rules	**Percentage of freelance rules (%)**	**AWPM**	**RWPM**	**RWPM-RT**	**I-RWPM**
80	0.02	1.315	1.476	0.293	0.289
90	0.03	2.699	2.739	0.460	0.476
100	0.03	5.299	4.122	0.766	0.616
110	0	10.396	7.785	1.141	1.129
120	0.02	20.431	12.036	1.615	1.577
130	0.02	39.078	20.028	2.633	2.655
140	0.05	57.169(99)	30.805	3.587	3.367
150	0.01	-	52.426	6.838	6.229
160	0.03	-	66.235(99)	7.259	7.162

compares AWPM, RWPM, RWPM-RT, and I-RWPM based on the average of 100 instances each, with a 300-second time limit per instance. The number in brackets indicates how many instances were successfully solved within the time limit by the corresponding method and is omitted if all 100 instances were solved. For a fixed number of rules (e.g., *#rules* = *N*), if any instances exceed the time limit, the solver is terminated, and "—" is marked for the corresponding encoding with *#rules* > *N*.

Table 11.2 clearly shows that RWPM-RT and I-RWPM significantly outperforms AWPM and RWPM in both solving time and the number of successfully solved instances. This demonstrates that RWPM-RT and I-RWPM provide a substantial advantage since they achieve more compact encoding of transitive laws.

Examining the average runtime of RWPM-RT and I-RWPM in Table 11.2, we observe that when the percentage of freelance rules in an MC-net is minimal, the advantage of I-RWPM over RWPM-RT is limited. A similar phenomenon can be seen in Figure 11.5a, where the percentage of freelance rules is only 5%. This suggests that, compared to RWPM-RT, I-RWPM's more compact encoding of rule relations and transitive laws (introduced in Section 11.3) is not comprehensively applied across all pairs of rules, resulting in only a limited improvement in performance. Readers may refer to Section 11.3.3 for specific examples illustrating the differences between the RWPM-RT and I-RWPM encodings.

11.5 Chapter Summary

This chapter has introduced an improved rule-based weighted partial MaxSAT (I-RWPM) encoding for solving the CSG problem, represented using the concise MC-net framework.

First, a specific set of freelance rules is identified, characterized by properties that allow these rules to be added to or removed from any feasible rule set without compromising its validity. The original MC-net problem is then decomposed into two subproblems involving nonoverlapping subsets: one containing all freelance rules and the other excluding them. Since solving the freelance rule subset is straightforward and requires no constraint encoding, the focus is solely on encoding the subproblem without freelance rules, effectively reducing the problem size and simplifying the overall solution process.

Second, a more compact encoding is employed to handle rule relations and transitive laws, which are fundamental for reasoning in this context. This enhanced encoding further reduces the number of variables and clauses.

Finally, the performance comparison shows that I-RWPM surpasses the existing three encodings, highlighting its superior efficiency, particularly in scenarios involving freelance rules in an MC-net.

References

1 X. Liao, M. Koshimura, K. Nomoto, S. Ueda, Y. Sakurai, and M. Yokoo. Improved WPM encoding for coalition structure generation under MC-nets. *Constraints*, 24(1):25–55, 2019.

2 N. Ohta, V. Conitzer, R. Ichimura, Y. Sakurai, A. Iwasaki, and M. Yokoo. Coalition structure generation utilizing compact characteristic function

representation. In *Proceedings of 15th International Conference on Principles and Practice of Constraint Programming (CP '09)*, 2009.

3 T. Michalak, D. Marciniak, M. Szamotulski, T. Rahwan, M. Wooldridge, P. McBurney, and N. Jennings. A logic-based representation for coalitional games with externalities. In *Proceedings of 9th International Conference on Autonomous Agents and Multiagent Systems*, 2010.

Appendix A

Complete File for Solving 4-Queens in DIMACS Format

The following shows the SAT encoding for solving 4-queens in DIMACS CNF input format.

```
c    SAT encoding for 4-queens
p    cnf 16 84
c    clauses generated by S1
1     2 3 4 0
−1    −2 0
−1    −3 0
−1    −4 0
−2    −3 0
−2    −4 0
−3    −4 0
5     6 7 8 0
−5    −6 0
−5    −7 0
−5    −8 0
−6    −7 0
−6    −8 0
−7    −8 0
9     10 11 12 0
−9    −10 0
−9    −11 0
−9    −12 0
−10   −11 0
−10   −12 0
−11   −12 0
13    14 15 16 0
−13   −14 0
```

Applied Satisfiability: Cryptography, Scheduling, and Coalitional Games, First Edition. Xiaojuan Liao and Miyuki Koshimura.

```
-13 -15 0
-13 -16 0
-14 -15 0
-14 -16 0
-15 -16 0
c clauses generated by S2
1 5 9 13 0
-1 -5 0
-1 -9 0
-1 -13 0
-5 -9 0
-5 -13 0
-9 -13 0
2 6 10 14 0
-2 -6 0
-2 -10 0
-2 -14 0
-6 -10 0
-6 -14 0
-10 -14 0
3 7 11 15 0
-3 -7 0
-3 -11 0
-3 -15 0
-7 -11 0
-7 -15 0
-11 -15 0
4 8 12 16 0
-4 -8 0
-4 -12 0
-4 -16 0
-8 -12 0
-8 -16 0
-12 -16 0
c clauses generated by S3
-3 -8 0
-2 -7 0
-2 -12 0
-7 -12 0
-1 -6 0
```

```
-1   -11 0
-1   -16 0
-6   -11 0
-6   -16 0
-11  -16 0
-5   -10 0
-5   -15 0
-10  -15 0
-9   -14 0
c    clauses generated by S4
-2   -5 0
-3   -6 0
-3   -9 0
-6   -9 0
-4   -7 0
-4   -10 0
-4   -13 0
-7   -10 0
-7   -13 0
-10  -13 0
-8   -11 0
-8   -14 0
-11  -14 0
-12  -15 0
```

Appendix B

A Sample of Sbox Expressed in ANF

Let $B_0 = \{b_0, b_1, b_2, b_3, b_4, b_5, b_6, b_7\}$, and $b_0b_1b_2b_4b_5b_6b_7$ be short for $b_0 \wedge b_1 \wedge b_2 \wedge b_4 \wedge b_5 \wedge b_6 \wedge b_7$, then $S_i(B_0)$ $(i = 0, \ldots, 7)$ is represented as the following eight functions in algebraic normal form (ANF).

$S_0(B_0) = b_0b_1b_2b_4b_5b_6b_7 \oplus b_0b_1b_2b_3b_4b_6b_7 \oplus b_0b_2b_3b_4b_5b_6b_7 \oplus$
$b_0b_2b_3b_4b_5b_6 \oplus b_1b_2b_3b_4b_6b_7 \oplus b_0b_1b_4b_5b_6b_7 \oplus b_0b_1b_2b_3b_5b_7 \oplus$
$b_0b_1b_2b_3b_6b_7 \oplus b_1b_3b_4b_5b_6b_7 \oplus b_0b_1b_2b_4b_5b_7 \oplus b_0b_1b_2b_3b_4b_7 \oplus$
$b_0b_1b_3b_4b_6b_7 \oplus b_0b_2b_3b_5b_6b_7 \oplus b_0b_1b_2b_5b_6b_7 \oplus b_2b_3b_4b_5b_6b_7 \oplus$
$b_1b_2b_4b_5b_6b_7 \oplus b_0b_1b_2b_3b_4b_6 \oplus b_0b_2b_3b_4b_5b_7 \oplus b_0b_3b_5b_6b_7 \oplus$
$b_0b_2b_3b_4b_5 \oplus b_2b_3b_4b_5b_7 \oplus b_1b_2b_4b_6b_7 \oplus b_0b_1b_5b_6b_7 \oplus b_0b_3b_4b_5b_7 \oplus$
$b_0b_1b_2b_3b_6 \oplus b_0b_1b_3b_5b_7 \oplus b_0b_1b_2b_6b_7 \oplus b_0b_2b_4b_6b_7 \oplus b_1b_2b_4b_5b_6 \oplus$
$b_0b_1b_4b_5b_6 \oplus b_0b_1b_2b_4b_7 \oplus b_0b_2b_3b_4b_6 \oplus b_0b_1b_2b_3b_4 \oplus b_0b_1b_2b_4b_5 \oplus$
$b_1b_2b_5b_6b_7 \oplus b_0b_1b_3b_5b_6 \oplus b_1b_2b_3b_5b_7 \oplus b_1b_3b_4b_5b_7 \oplus b_0b_3b_4b_5b_6 \oplus$
$b_0b_3b_4b_6b_7 \oplus b_1b_2b_3b_5b_6 \oplus b_0b_1b_2b_5b_7 \oplus b_0b_1b_3b_4b_6 \oplus b_2b_3b_4b_5b_6 \oplus$
$b_0b_1b_3b_6b_7 \oplus b_2b_4b_5b_6b_7 \oplus b_0b_2b_4b_5b_7 \oplus b_0b_1b_2b_3b_7 \oplus b_0b_2b_5b_7 \oplus$
$b_1b_2b_4b_6 \oplus b_0b_1b_2b_7 \oplus b_3b_5b_6b_7 \oplus b_1b_3b_6b_7 \oplus b_0b_2b_4b_5 \oplus b_0b_4b_5b_6 \oplus$
$b_1b_2b_3b_5 \oplus b_0b_1b_2b_3 \oplus b_0b_2b_3b_5 \oplus b_2b_3b_5b_7 \oplus b_0b_1b_4b_7 \oplus b_0b_1b_4b_5 \oplus$
$b_1b_3b_4b_6 \oplus b_2b_4b_5b_6 \oplus b_0b_1b_2b_5 \oplus b_2b_4b_5b_7 \oplus b_2b_3b_6b_7 \oplus b_1b_2b_3b_6 \oplus$
$b_0b_4b_5b_7 \oplus b_1b_5b_6b_7 \oplus b_1b_2b_4b_7 \oplus b_0b_3b_5b_6 \oplus b_0b_2b_3b_7 \oplus b_1b_2b_6b_7 \oplus$
$b_1b_2b_3b_7 \oplus b_1b_4b_5b_7 \oplus b_0b_4b_6b_7 \oplus b_0b_3b_4b_6 \oplus b_1b_4b_5b_6 \oplus b_0b_1b_2b_6 \oplus$
$b_0b_2b_5b_6 \oplus b_0b_2b_4b_7 \oplus b_1b_2b_3 \oplus b_3b_5b_7 \oplus b_0b_2b_4 \oplus b_3b_4b_7 \oplus b_1b_2b_6 \oplus$
$b_1b_4b_6 \oplus b_0b_4b_6 \oplus b_0b_1b_6 \oplus b_1b_3b_7 \oplus b_0b_3b_5 \oplus b_2b_3b_5 \oplus b_2b_5b_7 \oplus$
$b_2b_3b_7 \oplus b_0b_3b_6 \oplus b_0b_1b_4 \oplus b_5b_6b_7 \oplus b_1b_2b_4 \oplus b_0b_4b_7 \oplus b_0b_2b_5 \oplus$
$b_0b_1b_7 \oplus b_0b_3b_4 \oplus b_2b_5b_6 \oplus b_2b_6b_7 \oplus b_0b_2b_7 \oplus b_4b_5b_6 \oplus b_3b_6b_7 \oplus$
$b_1b_3b_4 \oplus b_1b_5b_6 \oplus b_0b_2b_6 \oplus b_2b_4b_7 \oplus b_1b_4 \oplus b_1b_6 \oplus b_0b_6 \oplus b_0b_1 \oplus$
$b_0b_5 \oplus b_6b_7 \oplus b_2b_7 \oplus b_2b_6 \oplus b_2b_4 \oplus b_5b_6 \oplus b_1b_3 \oplus b_1b_2 \oplus b_0b_4 \oplus$
$b_2b_3 \oplus b_4b_6 \oplus b_5b_7 \oplus b_0 \oplus b_2 \oplus b_3 \oplus \neg b_4.$

$S_1(B_0) = b_0b_1b_2b_3b_5b_6b_7 \oplus b_0b_1b_3b_4b_5b_6b_7 \oplus b_0b_1b_2b_4b_5b_6b_7 \oplus$
$b_0b_1b_2b_3b_4b_6b_7 \oplus b_0b_2b_3b_4b_6b_7 \oplus b_0b_1b_3b_4b_6b_7 \oplus b_1b_3b_4b_5b_6b_7 \oplus$
$b_0b_1b_2b_3b_5b_6 \oplus b_1b_2b_3b_4b_6b_7 \oplus b_0b_1b_3b_4b_5b_6 \oplus b_0b_1b_2b_4b_6b_7 \oplus$

Applied Satisfiability: Cryptography, Scheduling, and Coalitional Games,
First Edition. Xiaojuan Liao and Miyuki Koshimura.

$b_0b_1b_2b_4b_5b_7 \oplus b_1b_2b_3b_5b_6b_7 \oplus b_0b_1b_2b_3b_4b_6 \oplus b_1b_2b_3b_4b_5b_6 \oplus$
$b_0b_2b_3b_5b_6b_7 \oplus b_0b_2b_3b_4b_5b_7 \oplus b_0b_1b_2b_3b_4b_7 \oplus b_0b_1b_2b_5b_6b_7 \oplus$
$b_0b_2b_3b_4b_5b_6 \oplus b_1b_3b_4b_5b_6 \oplus b_1b_2b_3b_5b_6 \oplus b_0b_1b_2b_3b_6 \oplus b_0b_2b_5b_6b_7 \oplus$
$b_2b_3b_4b_5b_7 \oplus b_0b_1b_3b_4b_7 \oplus b_1b_2b_3b_4b_6 \oplus b_0b_2b_3b_5b_7 \oplus b_0b_1b_2b_6b_7 \oplus$
$b_0b_1b_2b_3b_4 \oplus b_0b_1b_3b_4b_6 \oplus b_0b_2b_4b_5b_6 \oplus b_0b_3b_5b_6b_7 \oplus b_0b_1b_5b_6b_7 \oplus$
$b_0b_1b_2b_4b_7 \oplus b_0b_1b_2b_3b_7 \oplus b_0b_1b_3b_5b_7 \oplus b_1b_2b_4b_5b_6 \oplus b_0b_2b_3b_6b_7 \oplus$
$b_2b_3b_4b_6b_7 \oplus b_0b_2b_3b_4b_7 \oplus b_0b_3b_4b_5b_7 \oplus b_0b_1b_2b_4b_6 \oplus b_1b_3b_4b_5b_7 \oplus$
$b_0b_1b_3b_6b_7 \oplus b_0b_1b_2b_4b_5 \oplus b_0b_1b_4b_5b_6 \oplus b_0b_1b_3b_5b_6 \oplus b_0b_4b_5b_6b_7 \oplus$
$b_2b_4b_5b_6b_7 \oplus b_0b_2b_3b_4b_5 \oplus b_0b_1b_4b_7 \oplus b_1b_3b_4b_7 \oplus b_0b_4b_5b_7 \oplus b_0b_2b_5b_6 \oplus$
$b_0b_3b_4b_6 \oplus b_1b_2b_6b_7 \oplus b_1b_3b_4b_5 \oplus b_1b_3b_5b_7 \oplus b_1b_2b_5b_7 \oplus b_2b_3b_5b_7 \oplus$
$b_0b_3b_4b_5 \oplus b_2b_3b_4b_5 \oplus b_1b_2b_3b_5 \oplus b_0b_1b_5b_6 \oplus b_2b_4b_5b_7 \oplus b_0b_1b_2b_6 \oplus$
$b_1b_4b_6b_7 \oplus b_0b_4b_5b_6 \oplus b_0b_1b_3b_6 \oplus b_3b_4b_6b_7 \oplus b_0b_2b_4b_5 \oplus b_1b_2b_4b_7 \oplus$
$b_0b_4b_6b_7 \oplus b_3b_4b_5b_6 \oplus b_1b_3b_5b_6 \oplus b_1b_4b_5b_7 \oplus b_0b_2b_3b_6 \oplus b_3b_4b_5b_7 \oplus$
$b_0b_1b_4b_5 \oplus b_0b_1b_3b_4 \oplus b_2b_3b_4b_7 \oplus b_2b_4b_6b_7 \oplus b_1b_2b_5b_6 \oplus b_3b_5b_6 \oplus$
$b_1b_3b_6 \oplus b_2b_3b_5 \oplus b_0b_5b_7 \oplus b_2b_6b_7 \oplus b_2b_3b_6 \oplus b_1b_3b_4 \oplus b_5b_6b_7 \oplus$
$b_3b_6b_7 \oplus b_1b_2b_4 \oplus b_0b_3b_4 \oplus b_0b_2b_3 \oplus b_3b_5b_7 \oplus b_3b_4b_7 \oplus b_0b_2b_7 \oplus$
$b_3b_4b_6 \oplus b_0b_1b_4 \oplus b_1b_2b_3 \oplus b_0b_1b_6 \oplus b_0b_6b_7 \oplus b_1b_3b_7 \oplus b_0b_4b_6 \oplus$
$b_2b_5b_7 \oplus b_2b_3b_4 \oplus b_1b_3b_5 \oplus b_1b_4b_7 \oplus b_0b_4b_7 \oplus b_0b_1b_3 \oplus b_1b_5b_6 \oplus$
$b_0b_4b_5 \oplus b_0b_4 \oplus b_3b_7 \oplus b_0b_7 \oplus b_0b_1 \oplus b_2b_3 \oplus b_4b_5 \oplus b_1b_3 \oplus b_2b_7 \oplus$
$b_0b_3 \oplus b_4b_6 \oplus b_1b_7 \oplus b_2b_6 \oplus b_1b_4 \oplus b_0b_2 \oplus b_7 \oplus b_6 \oplus b_3 \oplus \neg b_0.$

$S_2\left(B_0\right) = b_0b_2b_3b_4b_5b_6b_7 \oplus b_0b_1b_2b_3b_5b_6b_7 \oplus b_0b_1b_2b_3b_4b_6b_7 \oplus$
$b_0b_1b_3b_4b_5b_6b_7 \oplus b_1b_2b_3b_4b_5b_6b_7 \oplus b_0b_2b_3b_5b_6b_7 \oplus b_1b_2b_3b_4b_5b_7 \oplus$
$b_0b_1b_3b_5b_6b_7 \oplus b_0b_1b_2b_3b_4b_5 \oplus b_1b_2b_3b_4b_6b_7 \oplus b_0b_1b_2b_3b_5b_6 \oplus$
$b_0b_1b_3b_4b_6b_7 \oplus b_0b_1b_2b_4b_5b_7 \oplus b_0b_1b_2b_3b_4b_6 \oplus b_0b_1b_2b_3b_4b_7 \oplus$
$b_0b_2b_3b_4b_5b_6 \oplus b_0b_1b_2b_4b_5b_6 \oplus b_2b_3b_4b_5b_6b_7 \oplus b_0b_1b_4b_5b_6b_7 \oplus$
$b_0b_2b_3b_4b_6b_7 \oplus b_0b_2b_4b_5b_6b_7 \oplus b_0b_1b_3b_4b_5b_7 \oplus b_0b_3b_4b_5b_7 \oplus$
$b_1b_3b_4b_6b_7 \oplus b_0b_1b_2b_6b_7 \oplus b_0b_1b_4b_6b_7 \oplus b_1b_4b_5b_6b_7 \oplus b_1b_2b_4b_5b_6 \oplus$
$b_0b_2b_3b_4b_7 \oplus b_3b_4b_5b_6b_7 \oplus b_0b_1b_2b_3b_4 \oplus b_1b_2b_3b_4b_6 \oplus b_1b_3b_5b_6b_7 \oplus$
$b_0b_2b_3b_5b_7 \oplus b_1b_2b_3b_4b_5 \oplus b_1b_2b_4b_5b_7 \oplus b_1b_3b_4b_5b_6 \oplus b_0b_1b_3b_4b_7 \oplus$
$b_0b_2b_5b_6b_7 \oplus b_1b_2b_4b_6b_7 \oplus b_0b_1b_2b_3b_6 \oplus b_0b_1b_3b_6b_7 \oplus b_0b_1b_3b_4b_6 \oplus$
$b_0b_4b_5b_6b_7 \oplus b_0b_1b_2b_3b_5 \oplus b_0b_1b_2b_3b_7 \oplus b_0b_3b_5b_6b_7 \oplus b_1b_2b_3b_6b_7 \oplus$
$b_0b_2b_4b_5b_6 \oplus b_0b_1b_2b_5b_7 \oplus b_2b_3b_5b_6b_7 \oplus b_2b_3b_4b_6b_7 \oplus b_0b_1b_2b_4b_7 \oplus$
$b_0b_1b_2b_4b_5 \oplus b_0b_1b_3b_5b_7 \oplus b_0b_2b_3b_4 \oplus b_3b_5b_6b_7 \oplus b_1b_2b_3b_5 \oplus b_1b_2b_5b_7 \oplus$
$b_1b_3b_5b_7 \oplus b_0b_3b_5b_6 \oplus b_2b_4b_5b_7 \oplus b_0b_1b_3b_4 \oplus b_1b_5b_6b_7 \oplus b_0b_3b_4b_7 \oplus$
$b_3b_4b_5b_6 \oplus b_0b_2b_3b_7 \oplus b_0b_1b_2b_5 \oplus b_2b_4b_6b_7 \oplus b_0b_1b_3b_5 \oplus b_0b_2b_4b_7 \oplus$
$b_0b_1b_2b_7 \oplus b_0b_1b_5b_6 \oplus b_2b_5b_6b_7 \oplus b_4b_5b_6b_7 \oplus b_0b_3b_4b_6 \oplus b_1b_2b_6b_7 \oplus$
$b_2b_3b_4b_7 \oplus b_1b_2b_3b_4 \oplus b_0b_1b_3b_6 \oplus b_0b_2b_3b_5 \oplus b_1b_2b_3b_6 \oplus b_0b_1b_4b_6 \oplus$
$b_1b_4b_5b_7 \oplus b_2b_4b_5b_6 \oplus b_2b_3b_5b_6 \oplus b_0b_3b_5b_7 \oplus b_1b_3b_6b_7 \oplus b_1b_2b_4b_7 \oplus$
$b_0b_2b_4b_6 \oplus b_2b_3b_4b_6 \oplus b_0b_4b_6 \oplus b_2b_3b_4 \oplus b_0b_3b_6 \oplus b_1b_2b_5 \oplus b_0b_3b_7 \oplus$
$b_3b_5b_6 \oplus b_0b_2b_7 \oplus b_3b_4b_5 \oplus b_0b_1b_4 \oplus b_0b_6b_7 \oplus b_2b_4b_5 \oplus b_2b_3b_7 \oplus$
$b_1b_2b_7 \oplus b_1b_5b_7 \oplus b_0b_2b_5 \oplus b_0b_2b_4 \oplus b_0b_1b_7 \oplus b_1b_3b_5 \oplus b_1b_3b_4 \oplus$

$b_1b_4b_5 \oplus b_1b_2b_6 \oplus b_0b_3b_4 \oplus b_4b_6b_7 \oplus b_0b_5b_7 \oplus b_0b_4b_7 \oplus b_1b_2b_4 \oplus$
$b_0b_1b_3 \oplus b_0b_1b_5 \oplus b_1b_4b_7 \oplus b_3b_4b_6 \oplus b_1b_3b_7 \oplus b_0b_3b_5 \oplus b_2b_6b_7 \oplus$
$b_0b_2b_3 \oplus b_2b_3b_6 \oplus b_1b_5b_6 \oplus b_0b_4b_5 \oplus b_0b_1b_6 \oplus b_5b_6 \oplus b_0b_2 \oplus b_0b_4 \oplus$
$b_1b_4 \oplus b_6b_7 \oplus b_0b_5 \oplus b_2b_3 \oplus b_4b_7 \oplus b_0b_7 \oplus b_0b_6 \oplus b_3b_4 \oplus b_0b_3 \oplus$
$b_7 \oplus b_0 \oplus b_1 \oplus b_5.$

$S_3(B_0) = b_0b_1b_2b_3b_4b_5b_6 \oplus b_0b_1b_3b_4b_5b_6b_7 \oplus b_0b_2b_3b_4b_5b_6b_7 \oplus$
$b_0b_1b_2b_3b_5b_6b_7 \oplus b_0b_1b_2b_3b_6b_7 \oplus b_0b_1b_3b_4b_5b_7 \oplus b_0b_1b_2b_3b_4b_6 \oplus$
$b_1b_2b_3b_4b_5b_6 \oplus b_0b_1b_2b_4b_5b_7 \oplus b_1b_2b_3b_4b_5b_7 \oplus b_2b_3b_4b_5b_6b_7 \oplus$
$b_1b_2b_4b_5b_6b_7 \oplus b_0b_2b_3b_5b_6b_7 \oplus b_0b_1b_2b_3b_5b_7 \oplus b_0b_1b_2b_3b_5b_6 \oplus$
$b_1b_2b_3b_4b_6b_7 \oplus b_0b_1b_2b_4b_6b_7 \oplus b_1b_2b_3b_5b_6b_7 \oplus b_0b_2b_3b_4b_6b_7 \oplus$
$b_0b_1b_2b_3b_4b_7 \oplus b_0b_3b_4b_5b_6 \oplus b_0b_2b_3b_5b_7 \oplus b_0b_2b_4b_5b_7 \oplus$
$b_2b_3b_4b_5b_6 \oplus b_0b_4b_5b_6b_7 \oplus b_0b_1b_2b_3b_5 \oplus b_0b_3b_4b_6b_7 \oplus$
$b_0b_2b_5b_6b_7 \oplus b_0b_1b_2b_5b_6 \oplus b_2b_4b_5b_6b_7 \oplus b_0b_1b_2b_4b_7 \oplus b_0b_1b_5b_6b_7 \oplus$
$b_0b_1b_2b_3b_4 \oplus b_0b_1b_2b_3b_7 \oplus b_3b_4b_5b_6b_7 \oplus b_0b_1b_4b_5b_7 \oplus b_0b_2b_4b_5b_6 \oplus$
$b_0b_2b_3b_4b_5 \oplus b_1b_3b_4b_6b_7 \oplus b_0b_1b_3b_4b_7 \oplus b_1b_2b_3b_5b_6 \oplus b_0b_1b_2b_5b_7 \oplus$
$b_2b_3b_4b_5b_7 \oplus b_0b_1b_3b_4b_5 \oplus b_0b_1b_3b_5b_6 \oplus b_1b_2b_4b_5b_6 \oplus b_0b_1b_3b_4b_6 \oplus$
$b_1b_2b_5b_6b_7 \oplus b_0b_3b_4b_5b_7 \oplus b_0b_2b_3b_5b_6 \oplus b_0b_3b_5b_6b_7 \oplus b_1b_3b_5b_6b_7 \oplus$
$b_0b_2b_3b_6b_7 \oplus b_0b_3b_4b_6 \oplus b_0b_1b_2b_3 \oplus b_0b_4b_5b_6 \oplus b_0b_2b_6b_7 \oplus b_1b_2b_6b_7 \oplus$
$b_1b_3b_4b_6 \oplus b_4b_5b_6b_7 \oplus b_1b_2b_4b_5 \oplus b_1b_4b_5b_7 \oplus b_1b_2b_3b_7 \oplus b_1b_3b_5b_6 \oplus$
$b_2b_4b_5b_7 \oplus b_2b_3b_4b_5 \oplus b_1b_2b_3b_6 \oplus b_1b_4b_5b_6 \oplus b_1b_3b_4b_5 \oplus b_3b_4b_6b_7 \oplus$
$b_1b_5b_6b_7 \oplus b_0b_5b_6b_7 \oplus b_0b_1b_4b_6 \oplus b_1b_4b_6b_7 \oplus b_2b_4b_5b_6 \oplus b_0b_1b_2b_4 \oplus$
$b_1b_2b_3b_5 \oplus b_0b_1b_3b_6 \oplus b_3b_4b_5b_6 \oplus b_0b_1b_2b_5 \oplus b_0b_2b_4b_6 \oplus b_0b_1b_5b_6 \oplus$
$b_1b_2b_4b_7 \oplus b_0b_2b_3b_7 \oplus b_1b_3b_4b_7 \oplus b_1b_2b_5b_7 \oplus b_3b_5b_6b_7 \oplus b_2b_4b_6b_7 \oplus$
$b_0b_1b_6b_7 \oplus b_0b_2b_5b_6 \oplus b_0b_2b_4b_7 \oplus b_0b_3b_6b_7 \oplus b_0b_2b_7 \oplus b_0b_4b_5 \oplus$
$b_2b_3b_7 \oplus b_0b_2b_3 \oplus b_0b_1b_7 \oplus b_0b_1b_3 \oplus b_0b_2b_6 \oplus b_0b_3b_7 \oplus b_0b_3b_4 \oplus$
$b_0b_3b_6 \oplus b_0b_4b_6 \oplus b_3b_4b_7 \oplus b_1b_2b_6 \oplus b_0b_1b_4 \oplus b_1b_2b_7 \oplus b_1b_2b_5 \oplus$
$b_2b_4b_5 \oplus b_2b_3b_4 \oplus b_2b_5b_7 \oplus b_2b_3b_5 \oplus b_0b_1b_6 \oplus b_1b_3b_4 \oplus b_0b_1b_5 \oplus$
$b_1b_5b_6 \oplus b_0b_2b_4 \oplus b_5b_6b_7 \oplus b_3b_5b_6 \oplus b_1b_2b_3 \oplus b_3b_5b_7 \oplus b_6b_7 \oplus$
$b_2b_3 \oplus b_0b_3 \oplus b_5b_6 \oplus b_0b_7 \oplus b_3b_7 \oplus b_4b_5 \oplus b_1b_7 \oplus b_1b_2 \oplus b_5b_7 \oplus$
$b_3b_6 \oplus b_0 \oplus b_6 \oplus b_4 \oplus b_7.$

$S_4(B_0) = b_0b_1b_2b_3b_4b_5b_7 \oplus b_0b_1b_2b_3b_4b_6b_7 \oplus b_0b_2b_3b_4b_5b_6b_7 \oplus$
$b_0b_1b_2b_3b_4b_5 \oplus b_1b_2b_3b_4b_5b_6 \oplus b_0b_1b_2b_3b_4b_7 \oplus b_0b_1b_2b_3b_5b_6 \oplus$
$b_1b_2b_3b_5b_6b_7 \oplus b_0b_3b_4b_5b_6b_7 \oplus b_1b_2b_3b_4b_6b_7 \oplus b_1b_2b_3b_4b_5b_7 \oplus$
$b_0b_1b_3b_4b_5b_6 \oplus b_0b_1b_2b_4b_5b_6 \oplus b_0b_2b_3b_4b_5b_7 \oplus b_0b_1b_2b_3b_6 \oplus b_0b_1b_2b_3b_7 \oplus$
$b_0b_1b_2b_6b_7 \oplus b_0b_2b_3b_4b_6 \oplus b_1b_2b_3b_5b_7 \oplus b_0b_1b_3b_6b_7 \oplus b_0b_1b_3b_5b_7 \oplus$
$b_0b_3b_5b_6b_7 \oplus b_1b_2b_5b_6b_7 \oplus b_1b_4b_5b_6b_7 \oplus b_0b_1b_2b_5b_7 \oplus b_1b_2b_4b_5b_6 \oplus$
$b_0b_2b_3b_4b_5 \oplus b_0b_2b_3b_5b_6 \oplus b_0b_1b_5b_6b_7 \oplus b_1b_3b_4b_5b_6 \oplus b_2b_3b_4b_5b_7 \oplus$
$b_1b_2b_3b_5b_6 \oplus b_1b_3b_5b_6b_7 \oplus b_0b_1b_2b_4b_5 \oplus b_0b_1b_2b_3b_4 \oplus b_1b_3b_4b_5b_7 \oplus$
$b_0b_2b_4b_6b_7 \oplus b_0b_3b_4b_6b_7 \oplus b_2b_3b_4b_6b_7 \oplus b_1b_2b_4b_5b_7 \oplus b_0b_2b_3b_5b_7 \oplus$
$b_2b_4b_5b_6b_7 \oplus b_0b_1b_4b_5b_6 \oplus b_0b_1b_3b_4b_6 \oplus b_2b_3b_4b_5b_6 \oplus b_1b_2b_3b_4b_6 \oplus$
$b_3b_4b_5b_6b_7 \oplus b_0b_4b_5b_6b_7 \oplus b_0b_1b_6b_7 \oplus b_0b_3b_4b_5 \oplus b_3b_4b_6b_7 \oplus b_0b_2b_5b_7 \oplus$

$b_0b_3b_5b_6 \oplus b_0b_3b_5b_7 \oplus b_0b_1b_3b_5 \oplus b_0b_2b_3b_4 \oplus b_0b_1b_3b_6 \oplus b_0b_1b_4b_6 \oplus b_1b_3b_4b_7 \oplus b_0b_1b_2b_6 \oplus b_2b_3b_5b_7 \oplus b_1b_2b_4b_6 \oplus b_2b_3b_4b_7 \oplus b_0b_3b_4b_6 \oplus b_1b_2b_3b_5 \oplus b_0b_3b_6b_7 \oplus b_0b_1b_4b_5 \oplus b_1b_2b_5b_7 \oplus b_0b_1b_5b_6 \oplus b_2b_3b_5b_6 \oplus b_0b_4b_5b_6 \oplus b_1b_3b_4b_5 \oplus b_0b_4b_6b_7 \oplus b_0b_1b_5b_7 \oplus b_0b_1b_3b_7 \oplus b_3b_5b_6b_7 \oplus b_3b_4b_5b_7 \oplus b_1b_2b_3b_4 \oplus b_1b_2b_6b_7 \oplus b_0b_3b_4b_7 \oplus b_0b_3b_5 \oplus b_1b_4b_5 \oplus b_1b_2b_4 \oplus b_4b_6b_7 \oplus b_1b_3b_5 \oplus b_0b_3b_4 \oplus b_2b_3b_4 \oplus b_2b_3b_5 \oplus b_0b_2b_6 \oplus b_3b_4b_5 \oplus b_3b_4b_7 \oplus b_1b_5b_7 \oplus b_2b_4b_7 \oplus b_5b_6b_7 \oplus b_3b_5b_6 \oplus b_3b_4b_6 \oplus b_3b_6b_7 \oplus b_4b_5b_7 \oplus b_0b_2b_3 \oplus b_2b_4b_5 \oplus b_0b_4b_7 \oplus b_2b_5b_6 \oplus b_1b_6b_7 \oplus b_0b_4b_6 \oplus b_0b_4b_5 \oplus b_2b_3b_6 \oplus b_3b_5b_7 \oplus b_0b_6b_7 \oplus b_2b_6b_7 \oplus b_4b_5 \oplus b_2b_5 \oplus b_3b_4 \oplus b_6b_7 \oplus b_1b_6 \oplus b_0b_6 \oplus b_0b_7 \oplus b_1b_4 \oplus b_3b_7 \oplus b_0b_5 \oplus b_0b_4 \oplus b_3b_5 \oplus b_2b_3 \oplus b_5b_7 \oplus b_0b_1 \oplus b_1b_5 \oplus b_4b_6 \oplus b_2 \oplus b_3 \oplus b_0 \oplus b_5 \oplus b_1.$

$S_5(B_0) = b_0b_1b_2b_4b_5b_6b_7 \oplus b_0b_1b_2b_3b_5b_6b_7 \oplus b_0b_1b_2b_4b_5b_7 \oplus b_0b_1b_2b_3b_5b_7 \oplus b_1b_2b_4b_5b_6b_7 \oplus b_0b_2b_3b_4b_6b_7 \oplus b_0b_1b_2b_3b_5b_6 \oplus b_0b_2b_4b_5b_6b_7 \oplus b_1b_2b_3b_4b_6b_7 \oplus b_0b_1b_2b_4b_5b_6 \oplus b_0b_1b_2b_3b_4b_5 \oplus b_1b_2b_3b_5b_6b_7 \oplus b_0b_2b_4b_5b_7 \oplus b_0b_2b_3b_5b_7 \oplus b_0b_1b_5b_6b_7 \oplus b_1b_3b_4b_6b_7 \oplus b_0b_2b_3b_6b_7 \oplus b_0b_1b_4b_5b_7 \oplus b_0b_1b_2b_4b_6 \oplus b_0b_3b_4b_5b_6 \oplus b_2b_3b_5b_6b_7 \oplus b_0b_1b_2b_3b_5 \oplus b_0b_1b_2b_3b_7 \oplus b_1b_2b_4b_5b_6 \oplus b_0b_1b_4b_6b_7 \oplus b_0b_3b_4b_5b_7 \oplus b_0b_2b_3b_4b_7 \oplus b_0b_1b_2b_4b_5 \oplus b_0b_4b_5b_6b_7 \oplus b_2b_3b_4b_5b_6 \oplus b_1b_3b_4b_5b_6 \oplus b_0b_1b_2b_3b_6 \oplus b_0b_1b_4b_5b_6 \oplus b_0b_1b_2b_3b_4 \oplus b_1b_2b_3b_4b_6 \oplus b_2b_4b_5b_6b_7 \oplus b_0b_1b_2b_6b_7 \oplus b_0b_1b_3b_5b_7 \oplus b_1b_2b_3b_4b_7 \oplus b_3b_4b_5b_7 \oplus b_0b_1b_3b_4 \oplus b_1b_4b_5b_7 \oplus b_1b_2b_4b_7 \oplus b_1b_2b_3b_5 \oplus b_0b_4b_6b_7 \oplus b_2b_3b_4b_7 \oplus b_0b_1b_4b_6 \oplus b_0b_1b_5b_6 \oplus b_0b_1b_4b_5 \oplus b_0b_2b_4b_5 \oplus b_0b_2b_3b_4 \oplus b_3b_4b_6b_7 \oplus b_1b_3b_5b_6 \oplus b_1b_3b_4b_7 \oplus b_0b_1b_4b_7 \oplus b_0b_1b_5b_7 \oplus b_0b_1b_2b_3 \oplus b_3b_4b_5b_6 \oplus b_0b_1b_2b_5 \oplus b_0b_1b_3b_7 \oplus b_0b_3b_4b_7 \oplus b_2b_4b_5b_7 \oplus b_0b_2b_3b_5 \oplus b_1b_2b_4b_5 \oplus b_2b_3b_4b_6 \oplus b_1b_2b_3b_6 \oplus b_2b_3b_4b_5 \oplus b_1b_3b_4b_6 \oplus b_0b_5b_6b_7 \oplus b_2b_3b_5b_6 \oplus b_1b_3b_5b_7 \oplus b_0b_4b_7 \oplus b_4b_5b_7 \oplus b_2b_4b_5 \oplus b_0b_5b_6 \oplus b_0b_1b_6 \oplus b_5b_6b_7 \oplus b_0b_1b_3 \oplus b_0b_2b_6 \oplus b_1b_5b_7 \oplus b_2b_4b_7 \oplus b_2b_3b_7 \oplus b_1b_2b_5 \oplus b_1b_2b_4 \oplus b_0b_1b_2 \oplus b_1b_3b_5 \oplus b_0b_3b_5 \oplus b_1b_4b_6 \oplus b_0b_2b_4 \oplus b_2b_3b_6 \oplus b_1b_3b_6 \oplus b_2b_5b_7 \oplus b_3b_5b_7 \oplus b_1b_3b_7 \oplus b_2b_6b_7 \oplus b_0b_1b_5 \oplus b_1b_5b_6 \oplus b_1b_6b_7 \oplus b_1b_4b_7 \oplus b_0b_1b_7 \oplus b_0b_1b_4 \oplus b_3b_4b_5 \oplus b_1b_2b_6 \oplus b_3b_4 \oplus b_1b_6 \oplus b_2b_4 \oplus b_1b_5 \oplus b_4b_6 \oplus b_5b_7 \oplus b_0b_3 \oplus b_7 \oplus b_6 \oplus \neg b_4.$

$S_6(B_0) = b_0b_1b_3b_4b_5b_6b_7 \oplus b_0b_1b_2b_4b_5b_6b_7 \oplus b_0b_1b_3b_4b_5b_7 \oplus b_0b_1b_2b_3b_5b_6 \oplus b_0b_1b_2b_3b_4b_7 \oplus b_0b_1b_3b_4b_5b_6 \oplus b_0b_1b_2b_4b_6b_7 \oplus b_0b_1b_2b_4b_5b_7 \oplus b_0b_1b_3b_5b_6b_7 \oplus b_0b_1b_3b_4b_6b_7 \oplus b_1b_3b_4b_5b_6b_7 \oplus b_0b_1b_2b_4b_5b_6 \oplus b_0b_1b_2b_5b_6b_7 \oplus b_0b_4b_5b_6b_7 \oplus b_1b_2b_3b_5b_7 \oplus b_1b_2b_3b_4b_5 \oplus b_0b_1b_3b_6b_7 \oplus b_0b_1b_4b_5b_7 \oplus b_0b_1b_2b_3b_5 \oplus b_1b_4b_5b_6b_7 \oplus b_0b_3b_4b_6b_7 \oplus b_0b_2b_5b_6b_7 \oplus b_0b_1b_2b_3b_7 \oplus b_1b_2b_4b_5b_6 \oplus b_0b_2b_4b_5b_7 \oplus b_2b_3b_4b_6b_7 \oplus b_0b_1b_2b_3b_6 \oplus b_1b_3b_4b_5b_6 \oplus b_1b_2b_5b_6b_7 \oplus b_2b_3b_5b_6b_7 \oplus b_1b_2b_3b_4b_7 \oplus b_3b_4b_5b_6b_7 \oplus b_0b_2b_3b_5b_6 \oplus b_1b_3b_4b_5b_7 \oplus b_0b_1b_3b_4b_5 \oplus b_0b_2b_4b_6b_7 \oplus b_0b_2b_3b_4b_5 \oplus b_0b_1b_3b_6 \oplus b_1b_2b_4b_7 \oplus b_1b_2b_3b_5 \oplus b_0b_3b_4b_7 \oplus b_0b_1b_6b_7 \oplus$

$b_4b_5b_6b_7 \oplus b_1b_3b_5b_7 \oplus b_0b_2b_4b_6 \oplus b_0b_2b_3b_5 \oplus b_1b_2b_3b_4 \oplus b_0b_3b_4b_6 \oplus$
$b_2b_3b_4b_5 \oplus b_2b_3b_5b_6 \oplus b_1b_2b_3b_6 \oplus b_1b_4b_6b_7 \oplus b_0b_1b_2b_5 \oplus b_0b_1b_3b_4 \oplus$
$b_0b_2b_3b_7 \oplus b_1b_4b_5b_7 \oplus b_1b_3b_4b_6 \oplus b_1b_3b_4b_7 \oplus b_0b_2b_4b_5 \oplus b_2b_3b_4b_6 \oplus$
$b_1b_2b_4b_5 \oplus b_0b_2b_3b_6 \oplus b_0b_2b_5b_7 \oplus b_2b_3b_5b_7 \oplus b_0b_1b_3b_7 \oplus b_0b_4b_6b_7 \oplus$
$b_0b_1b_2b_4 \oplus b_3b_4b_6 \oplus b_0b_2b_5 \oplus b_0b_4b_7 \oplus b_1b_5b_6 \oplus b_0b_1b_5 \oplus b_1b_2b_5 \oplus$
$b_1b_3b_6 \oplus b_0b_3b_5 \oplus b_5b_6b_7 \oplus b_0b_1b_4 \oplus b_0b_4b_5 \oplus b_1b_6b_7 \oplus b_0b_1b_3 \oplus$
$b_0b_2b_4 \oplus b_0b_2b_6 \oplus b_3b_5b_7 \oplus b_2b_4b_6 \oplus b_0b_5b_6 \oplus b_4b_5b_6 \oplus b_0b_4b_6 \oplus$
$b_2b_3b_7 \oplus b_1b_3b_4 \oplus b_1b_4b_7 \oplus b_1b_2b_7 \oplus b_2b_4b_7 \oplus b_3b_6b_7 \oplus b_2b_3b_4 \oplus$
$b_0b_5b_7 \oplus b_0b_3b_6 \oplus b_1b_4b_6 \oplus b_1b_2b_6 \oplus b_3b_5 \oplus b_0b_5 \oplus b_2b_3 \oplus b_5b_7 \oplus$
$b_1b_7 \oplus b_1b_3 \oplus b_4b_6 \oplus b_3b_7 \oplus b_0b_4 \oplus b_0b_7 \oplus b_5 \oplus b_6 \oplus \neg b_3.$

$S_7\left(B_0\right) = b_0b_2b_3b_4b_5b_6b_7 \oplus b_0b_1b_3b_4b_5b_6b_7 \oplus b_0b_2b_4b_5b_6b_7 \oplus$
$b_0b_2b_3b_4b_6b_7 \oplus b_0b_2b_3b_4b_5b_7 \oplus b_0b_1b_3b_4b_5b_7 \oplus b_0b_1b_2b_4b_5b_7 \oplus$
$b_1b_3b_4b_5b_6b_7 \oplus b_0b_1b_3b_5b_6b_7 \oplus b_0b_1b_2b_3b_6b_7 \oplus b_0b_1b_3b_4b_6b_7 \oplus$
$b_0b_2b_3b_5b_6b_7 \oplus b_0b_2b_3b_4b_5b_6 \oplus b_0b_1b_3b_4b_7 \oplus b_0b_3b_5b_6b_7 \oplus$
$b_0b_1b_2b_3b_6 \oplus b_0b_2b_3b_4b_6 \oplus b_1b_2b_3b_5b_6 \oplus b_0b_1b_2b_3b_4 \oplus b_1b_4b_5b_6b_7 \oplus$
$b_0b_1b_3b_4b_5 \oplus b_0b_1b_2b_5b_7 \oplus b_0b_2b_3b_4b_5 \oplus b_2b_3b_4b_5b_6 \oplus b_0b_1b_5b_6b_7 \oplus$
$b_0b_1b_2b_4b_6 \oplus b_0b_2b_3b_4b_7 \oplus b_0b_1b_4b_5b_7 \oplus b_0b_3b_4b_6b_7 \oplus b_1b_2b_4b_5b_6 \oplus$
$b_3b_4b_5b_6b_7 \oplus b_0b_1b_4b_5b_6 \oplus b_0b_3b_4b_5b_6 \oplus b_1b_2b_5b_6b_7 \oplus b_1b_2b_4b_6b_7 \oplus$
$b_0b_1b_2b_6b_7 \oplus b_1b_3b_5b_6b_7 \oplus b_0b_1b_4b_7 \oplus b_0b_1b_2b_4 \oplus b_2b_4b_6b_7 \oplus b_1b_3b_4b_7 \oplus$
$b_1b_4b_5b_7 \oplus b_4b_5b_6b_7 \oplus b_1b_2b_4b_6 \oplus b_2b_5b_6b_7 \oplus b_1b_5b_6b_7 \oplus b_1b_2b_3b_5 \oplus$
$b_1b_2b_3b_7 \oplus b_0b_2b_4b_7 \oplus b_0b_2b_3b_5 \oplus b_0b_1b_2b_7 \oplus b_0b_3b_6b_7 \oplus b_0b_3b_4b_7 \oplus$
$b_0b_1b_3b_6 \oplus b_0b_1b_2b_3 \oplus b_0b_3b_4b_6 \oplus b_3b_4b_5b_6 \oplus b_0b_3b_5b_6 \oplus b_1b_2b_3b_4 \oplus$
$b_0b_2b_3b_7 \oplus b_0b_4b_6b_7 \oplus b_1b_4b_6b_7 \oplus b_0b_3b_5b_7 \oplus b_1b_2b_4b_5 \oplus b_0b_2b_3b_6 \oplus$
$b_2b_3b_4b_7 \oplus b_1b_3b_6b_7 \oplus b_0b_1b_2b_5 \oplus b_0b_2b_4b_6 \oplus b_0b_1b_3b_4 \oplus b_1b_3b_6 \oplus$
$b_2b_4b_6 \oplus b_1b_2b_6 \oplus b_0b_2b_7 \oplus b_1b_4b_7 \oplus b_0b_3b_6 \oplus b_0b_3b_7 \oplus b_2b_6b_7 \oplus$
$b_1b_2b_3 \oplus b_0b_4b_5 \oplus b_2b_5b_6 \oplus b_0b_1b_5 \oplus b_0b_6b_7 \oplus b_0b_1b_6 \oplus b_4b_5b_7 \oplus$
$b_0b_2b_3 \oplus b_0b_2b_5 \oplus b_3b_5b_7 \oplus b_0b_5b_6 \oplus b_0b_1b_4 \oplus b_0b_3b_5 \oplus b_3b_4b_5 \oplus$
$b_4b_5b_6 \oplus b_2b_3b_5 \oplus b_1b_3b_5 \oplus b_4b_6b_7 \oplus b_0b_6 \oplus b_2b_4 \oplus b_2b_6 \oplus b_3b_5 \oplus$
$b_1b_7 \oplus b_0b_2 \oplus b_5b_7 \oplus b_4b_6 \oplus b_1b_2 \oplus b_0b_7 \oplus b_7 \oplus b_4 \oplus b_5 \oplus b_2.$

Appendix C

Complete File Generated by MaxSAT for Solving Overloaded Scheduling in WPM Input Format

The following shows the file in DIMACS CNF input format generated by MaxSAT for solving the example in Section 6.2.3.

```
p	wcnf 14 23 7
c	soft
1	9 0
2	10 0
3	11 0
c
c	constraint (C1)
7	1 0
7	5 0
7	7 0
c
c	constraint (C2)
7	12 0
c
c	constraint (C3)
7	13 14 0
c
c	constraint (C4)
7	10 2 0
7	10 8 0
c
c	constraint (C5)
7	−2 1 0
7	−4 3 0
7	−6 5 0
7	−8 7 0
```

Applied Satisfiability: Cryptography, Scheduling, and Coalitional Games, First Edition. Xiaojuan Liao and Miyuki Koshimura.

```
c
c   constraint (C6)
7   −4 −9 0
7   −6 −10 0
7   −8 −11 0
c
c   constraint (C7)
7   −1 −12 3 0
7   −2 −12 4 0
7   −1 −13 8 0
7   −2 −13 8 0
7   −7 −14 2 0
7   −8 −14 2 0
```

Appendix D

Complete File Generated by RWPM for Example 8.9 in WPM Input Format

The following shows the file generated by RWPM for solving Example 8.9 in DIMACS CNF input format.

```
c    #agent: 4 #rule: 4
c    the maximum cost: 8
p    wcnf 10 25 9
c
c    soft clause
2    1 0
2    2 0
1    3 0
3    4 0
c
c    hard clauses generated from compS
9    −1 −2 5 0
9    −5 1 0
9    −5 2 0
9    −1 −3 6 0
9    −6 1 0
9    −6 3 0
c
c    hard clauses generated from compD
9    −1 −4 −7 0
9    −2 −3 −8 0
c
c    hard clauses generated from incomp
9    −2 −4 0
```

```
c
c    transitive laws
9    −5 −8 6 0
9    −5 −9 7 0
9    −5 −6 8 0
9    −5 −7 9 0
9    −6 −8 5 0
9    −7 −9 5 0
9    −6 −10 7 0
9    −6 −7 10 0
9    −7 −10 6 0
9    −8 −10 9 0
9    −8 −9 10 0
9    −9 −10 8 0
```

Appendix E

Complete File Generated by RWPM for Example 8.11 in WPM Input Format

The following shows the file generated by RWPM for solving Example 8.11 in DIMACS CNF input format.

```
c   #agent: 4 #rule: 6
c   the maximum cost: 5
p   wcnf 21 47 6
c
c   soft clause
2   1 0
1   2 0
1   3 0
1   4 0
c
c   hard clauses generated from positive rules with externalities
6   −4 5 0
6   −4 6 0
c
c   hard clauses generated from compS
6   −1 −5 10 0
6   −10 1 0
6   −10 5 0
6   −3 −6 18 0
6   −18 3 0
6   −18 6 0
c
c   hard clauses generated from compD
6   −1 −3 −8 0
6   −2 −3 −12 0
6   −2 −6 −15 0
```

Applied Satisfiability: Cryptography, Scheduling, and Coalitional Games, First Edition. Xiaojuan Liao and Miyuki Koshimura.

```
6  −3 −5 −17 0
6  −5 −6 −21 0
c
c  hard clauses generated from incomp
c
c  transitive laws
6  −7 −12 8 0
6  −7 −14 10 0
6  −7 −15 11 0
6  −7 −8 12 0
6  −7 −10 14 0
6  −7 −11 15 0
6  −8 −12 7 0
6  −10 −14 7 0
6  −11 −15 7 0
6  −8 −17 10 0
6  −8 −18 11 0
6  −8 −10 17 0
6  −8 −11 18 0
6  −10 −17 8 0
6  −11 −18 8 0
6  −10 −21 11 0
6  −10 −11 21 0
6  −11 −21 10 0
6  −12 −17 14 0
6  −12 −18 15 0
6  −12 −14 17 0
6  −12 −15 18 0
6  −14 −17 12 0
6  −15 −18 12 0
6  −14 −21 15 0
6  −14 −15 21 0
6  −15 −21 14 0
6  −17 −21 18 0
6  −17 −18 21 0
6  −18 −21 17 0
```

Appendix F

Complete File Generated by RWPM for Example 8.12 in WPM Input Format

The following shows the file generated by RWPM for solving Example 8.12 in DIMACS CNF input format.

```
c   #agent: 4 #rule: 5
c   the maximum cost: 7
p   wcnf 15 50 8
c
c   soft clause
1   −1 0
4   2 0
2   3 0
c
c   hard clause generated from dummy rules
8   1 4 5 0
c
c   hard clauses generated from compS
8   −1 −2 6 0
8   −6 1 0
8   −6 2 0
8   −3 −5 14 0
8   −14 3 0
8   −14 5 0
8   −4 −5 15 0
8   −15 4 0
8   −15 5 0
c
c   hard clauses generated from compD
8   −3 −4 −13 0
c
```

Applied Satisfiability: Cryptography, Scheduling, and Coalitional Games,
First Edition. Xiaojuan Liao and Miyuki Koshimura.

```
c  hard clauses generated from incomp
8  -1 -3 0
8  -1 -4 0
8  -1 -5 0
8  -2 -3 0
8  -2 -4 0
8  -2 -5 0
c
c  transitive laws
8  -6 -10 7 0
8  -6 -11 8 0
8  -6 -12 9 0
8  -6 -7 10 0
8  -6 -8 11 0
8  -6 -9 12 0
8  -7 -10 6 0
8  -8 -11 6 0
8  -9 -12 6 0
8  -7 -13 8 0
8  -7 -14 9 0
8  -7 -8 13 0
8  -7 -9 14 0
8  -8 -13 7 0
8  -9 -14 7 0
8  -8 -15 9 0
8  -8 -9 15 0
8  -9 -15 8 0
8  -10 -13 11 0
8  -10 -14 12 0
8  -10 -11 13 0
8  -10 -12 14 0
8  -11 -13 10 0
8  -12 -14 10 0
8  -11 -15 12 0
8  -11 -12 15 0
8  -12 -15 11 0
8  -13 -15 14 0
8  -13 -14 15 0
8  -14 -15 13 0
```

Appendix G

Complete File Generated by RWPM for Example 8.13 in WPM Input Format

The following shows the file generated by RWPM for solving Example 8.13 in DIMACS CNF input format.

```
c   #agent: 4 #rule: 8
c   the maximum cost: 5
p   wcnf 36 143 6
c
c   soft clause
2   1 0
1   2 0
1   −3 0
1   −4 0
c
c   hard clauses generated from dummy rules without ex
6   3 5 0
c
c   hard clauses generated from negative rules with ex
6   4 6 7 8 0
c
c   hard clauses generated from compS
6   −1 −5 12 0
6   −12 1 0
6   −12 5 0
6   −1 −7 14 0
6   −14 1 0
6   −14 7 0
6   −3 −8 26 0
6   −26 3 0
6   −26 8 0
```

Applied Satisfiability: Cryptography, Scheduling, and Coalitional Games, First Edition. Xiaojuan Liao and Miyuki Koshimura.

```
6   -5 -6 31 0
6   -31 5 0
6   -31 6 0
6   -5 -7 32 0
6   -32 5 0
6   -32 7 0
6   -5 -8 33 0
6   -33 5 0
6   -33 8 0
6   -6 -7 34 0
6   -34 6 0
6   -34 7 0
6   -7 -8 36 0
6   -36 7 0
6   -36 8 0
c
c   hard clauses generated from compD
6   -1 -3 -10 0
6   -2 -3 -16 0
6   -2 -5 -18 0
6   -2 -7 -20 0
6   -3 -6 -24 0
c
c   hard clauses generated from incomp
6   -1 -6 0
6   -2 -8 0
6   -3 -7 0
c
c   transitive laws
6   -9 -16 10 0
6   -9 -18 12 0
6   -9 -19 13 0
6   -9 -20 14 0
6   -9 -21 15 0
6   -9 -10 16 0
6   -9 -12 18 0
6   -9 -13 19 0
6   -9 -14 20 0
6   -9 -15 21 0
6   -10 -16 9 0
6   -12 -18 9 0
```

6 −13 −19 9 0
6 −14 −20 9 0
6 −15 −21 9 0
6 −10 −23 12 0
6 −10 −24 13 0
6 −10 −25 14 0
6 −10 −26 15 0
6 −10 −12 23 0
6 −10 −13 24 0
6 −10 −14 25 0
6 −10 −15 26 0
6 −12 −23 10 0
6 −13 −24 10 0
6 −14 −25 10 0
6 −15 −26 10 0
6 −12 −31 13 0
6 −12 −32 14 0
6 −12 −33 15 0
6 −12 −13 31 0
6 −12 −14 32 0
6 −12 −15 33 0
6 −13 −31 12 0
6 −14 −32 12 0
6 −15 −33 12 0
6 −13 −34 14 0
6 −13 −35 15 0
6 −13 −14 34 0
6 −13 −15 35 0
6 −14 −34 13 0
6 −15 −35 13 0
6 −14 −36 15 0
6 −14 −15 36 0
6 −15 −36 14 0
6 −16 −23 18 0
6 −16 −24 19 0
6 −16 −25 20 0
6 −16 −26 21 0
6 −16 −18 23 0
6 −16 −19 24 0
6 −16 −20 25 0
6 −16 −21 26 0

```
6	–18 –23 16 0
6	–19 –24 16 0
6	–20 –25 16 0
6	–21 –26 16 0
6	–18 –31 19 0
6	–18 –32 20 0
6	–18 –33 21 0
6	–18 –19 31 0
6	–18 –20 32 0
6	–18 –21 33 0
6	–19 –31 18 0
6	–20 –32 18 0
6	–21 –33 18 0
6	–19 –34 20 0
6	–19 –35 21 0
6	–19 –20 34 0
6	–19 –21 35 0
6	–20 –34 19 0
6	–21 –35 19 0
6	–20 –36 21 0
6	–20 –21 36 0
6	–21 –36 20 0
6	–23 –31 24 0
6	–23 –32 25 0
6	–23 –33 26 0
6	–23 –24 31 0
6	–23 –25 32 0
6	–23 –26 33 0
6	–24 –31 23 0
6	–25 –32 23 0
6	–26 –33 23 0
6	–24 –34 25 0
6	–24 –35 26 0
6	–24 –25 34 0
6	–24 –26 35 0
6	–25 –34 24 0
6	–26 –35 24 0
6	–25 –36 26 0
6	–25 –26 36 0
6	–26 –36 25 0
6	–31 –34 32 0
```

```
6    −31 −35 33 0
6    −31 −32 34 0
6    −31 −33 35 0
6    −32 −34 31 0
6    −33 −35 31 0
6    −32 −36 33 0
6    −32 −33 36 0
6    −33 −36 32 0
6    −34 −36 35 0
6    −34 −35 36 0
6    −35 −36 34 0
```

Appendix H

Complete File Generated by AWPM for Example 9.2 in WPM Input Format

The following shows the file generated by AWPM for solving Example 9.2 in DIMACS CNF input format.

```
c  #agent: 4 #rule: 4
c  the maximum cost: 8
p  wcnf 10 21 9
c
c  soft clauses and their derived hard clauses
2  7 0
9  −7 1 0
c
2  8 0
9  −8 4 0
9  −8 −5 0
c
1  9 0
9  −9 3 0
c
3  10 0
9  −10 −4 0
c
c  transitive laws
9  −1 −2 4 0
9  −1 −3 5 0
9  −1 −4 2 0
9  −1 −5 3 0
9  −2 −3 6 0
9  −2 −6 3 0
9  −2 −4 1 0
```

Applied Satisfiability: Cryptography, Scheduling, and Coalitional Games,
First Edition. Xiaojuan Liao and Miyuki Koshimura.

```
9       -3 -5 1 0
9       -3 -6 2 0
9       -4 -5 6 0
9       -4 -6 5 0
9       -5 -6 4 0
```

Appendix I

Complete File Generated by AWPM for Example 9.3 in WPM Input Format

The following shows the file generated by AWPM for solving Example 9.3 in DIMACS CNF input format.

```
c    #agent: 4 #rule: 4
c    the maximum cost: 5
p    wcnf 10 22 6
c
c    soft clauses and their derived hard clauses
2    7 0
6    −7 1 0
c
1    8 0
6    −8 −6 0
c
1    9 0
6    −9 −3 0
c
1    10 0
6    −10 1 0
6    −10 −3 0
6    −10 −6 0
c
c    transitive laws
6    −1 −2 4 0
6    −1 −3 5 0
6    −1 −4 2 0
6    −1 −5 3 0
6    −2 −3 6 0
6    −2 −6 3 0
```

Applied Satisfiability: Cryptography, Scheduling, and Coalitional Games,
First Edition. Xiaojuan Liao and Miyuki Koshimura.

```
6    −2 −4 1 0
6    −3 −5 1 0
6    −3 −6 2 0
6    −4 −5 6 0
6    −4 −6 5 0
6    −5 −6 4 0
```

Appendix J

Complete File Generated by AWPM for Example 9.4 in WPM Input Format

The following shows the file generated by AWPM for solving Example 9.4 in DIMACS CNF input format.

```
c  #agent: 4 #rule: 3
c  the maximum cost: 7
p  wcnf 9 20 8
c
c  soft clauses and their derived hard clauses
1  −7 0
8  7 −1 3 0
c
4  8 0
8  −8 1 0
8  −8 2 0
8  −8 −3 0
c
2  9 0
8  −9 5 0
c
c  transitive laws
8  −1 −2 4 0
8  −1 −3 5 0
8  −1 −4 2 0
8  −1 −5 3 0
8  −2 −3 6 0
8  −2 −6 3 0
8  −2 −4 1 0
```

Applied Satisfiability: Cryptography, Scheduling, and Coalitional Games,
First Edition. Xiaojuan Liao and Miyuki Koshimura.

```
8      −3 −5 1 0
8      −3 −6 2 0
8      −4 −5 6 0
8      −4 −6 5 0
8      −5 −6 4 0
```

Appendix K

Complete File Generated by AWPM for Example 9.5 in WPM Input Format

The following shows the file generated by AWPM for solving Example 9.5 in DIMACS CNF input format.

```
c    #agent: 4 #rule: 4
c    the maximum cost: 3
p    wcnf 10 20 6
c
c    soft clauses and their derived hard clauses
2    7 0
6    −7 1 0
c
1    8 0
6    −8 −6 0
c
1    −9 0
6    9 3 0
c
1    −10 0
6    10 −1 3 2 0
c
c    transitive laws
6    −1 −2 4 0
6    −1 −3 5 0
6    −1 −4 2 0
6    −1 −5 3 0
6    −2 −3 6 0
6    −2 −6 3 0
```

Applied Satisfiability: Cryptography, Scheduling, and Coalitional Games,
First Edition. Xiaojuan Liao and Miyuki Koshimura.

```
6    −2 −4 1 0
6    −3 −5 1 0
6    −3 −6 2 0
6    −4 −5 6 0
6    −4 −6 5 0
6    −5 −6 4 0
```

Appendix L

Proof of Formula in Lemma 10.3

This part proves $F = C_m^2 \left(1 - \frac{C_{m-\overline{m}}^{\overline{m}}}{C_m^{\overline{m}}}\right) - m\left(\overline{m} - 1\right) > 0$, where $\overline{m} \leq m/2$.

Proof: Given $m, k \in \mathbb{Z}^+$ and $k \leq m/2$, Eq. 1 $= \frac{P_{m-1}^{k-1}}{P_{m-k}^{k-1}} \geq 1$ obviously holds, where P_m^k means the number of k-permutations of n. Formally, $P_m^k = \frac{m!}{(m-k)!}$. Next, we transform Eq. 1 as follows:

$$
\begin{aligned}
\text{Eq. 1} \Rightarrow & \frac{mP_{m-1}^{k-1}}{P_{m-k}^{k-1}} \geq m \\
\Rightarrow & \frac{P_m^k}{P_{m-k}^{k-1}} \geq m > m - 1 \\
\Rightarrow & \frac{m!}{(m-k)!} > \frac{(m-1)(m-k)!}{(m-2k+1)!} \\
\Rightarrow & m!\,(m-2k+1)! > (m-1)(m-k)\,!^2 \\
\Rightarrow & m!\,(m-2k+1)(m-2k)! > (m-1)(m-k)\,!^2 \\
\Rightarrow & m - 2k + 1 > \frac{(m-1)(m-k)\,!^2}{m!\,(m-2k)!} \\
\Rightarrow & \frac{(m-1) - 2(k-1)}{m-1} > \frac{(m-k)\,!^2}{m!\,(m-2k)!} \\
\Rightarrow & 1 - \frac{(m-k)\,!^2}{m!\,(m-2k)!} > \frac{2(k-1)}{m-1} \\
\Rightarrow & 1 - \frac{(m-k)!/(m-2k)!k!}{m!/(m-k)!k!} > \frac{2m(m-2)!\,(k-1)}{m!} \\
\Rightarrow & 1 - \frac{C_{m-k}^k}{C_m^k} > \frac{2m(m-2)!\,(k-1)}{m!}
\end{aligned}
$$

Applied Satisfiability: Cryptography, Scheduling, and Coalitional Games,
First Edition. Xiaojuan Liao and Miyuki Koshimura.

$$\Rightarrow \frac{m!}{2!\,(m-2)!}\left(1-\frac{C_{m-k}^{k}}{C_{m}^{k}}\right) > m\,(k-1)$$

$$\Rightarrow C_{m}^{2}\left(1-\frac{C_{m-k}^{k}}{C_{m}^{k}}\right) > m\,(k-1)$$

Using $\overline{m}$ to replace k, we have $F = C_{m}^{2} \cdot \left(1-\frac{C_{m-\overline{m}}^{\overline{m}}}{C_{m}^{\overline{m}}}\right) - m\left(\overline{m}-1\right) > 0$.

Appendix M

Calculation of $\overline{m}$ in Chapter 10

Let random variable X represent the number of agents contained in a condition, $X = \{1, 2, \ldots, m\}$. Based on a method of generating instances introduced in Chapter 8, each condition is created with one random agent, and then a new random agent is repeatedly added with the probability of α until an agent is not added or the condition includes all agents. Based on this method, the probability that $X = t$ is represented as follows:

$$P(X) = \begin{cases} \alpha^{t-1}(1-\alpha) & \text{if } t < m, \\ \alpha^{m-1} & \text{otherwise} \end{cases}$$

Then the expectation of X is $E(X) = \sum_{t=1}^{m-1} t\alpha^{t-1}(1-\alpha) + m\alpha^{m-1}$. By simplifying the formula, we have

$$E(X) = \frac{1-\alpha^m}{1-\alpha}$$

Thus, $\overline{m}$ can be estimated by the expected value of X, i.e., $\overline{m} = \frac{1-\alpha^m}{1-\alpha}$.[1]

1 More precisely, depending on the number of agents and the value of α, the average number $\overline{m}$ will either be $\lfloor \frac{1-\alpha^m}{1-\alpha} \rfloor$ or $\lceil \frac{1-\alpha^m}{1-\alpha} \rceil$. However, to simplify the discussion without a loss of generality, we assume that $\overline{m}$ is just $\frac{1-\alpha^m}{1-\alpha}$.

Applied Satisfiability: Cryptography, Scheduling, and Coalitional Games, First Edition. Xiaojuan Liao and Miyuki Koshimura.

Appendix N

Complete File Generated by RWPM-RT for Example 10.3 in WPM Input Format

The following shows the file generated by RWPM-RT for solving Example 10.3 in DIMACS CNF input format.

```
c    #agent: 4 #rule: 4
c    the maximum cost: 8
p    wcnf 8 14 9
c
c    soft clause
2    1 0
2    2 0
1    3 0
3    4 0
c
c    hard clauses generated from compS
9    −1 −2 5 0
9    −5 1 0
9    −5 2 0
9    −1 −3 6 0
9    −6 1 0
9    −6 3 0
c
c    hard clauses generated from compD
9    −1 −4 −7 0
9    −2 −3 −8 0
c
c    hard clauses generated from incomp
9    −2 −4 0
c
c    transitive laws
9    −5 −6 8 0
```

Applied Satisfiability: Cryptography, Scheduling, and Coalitional Games, First Edition. Xiaojuan Liao and Miyuki Koshimura.

Appendix O

Complete File Generated by RWPM-RT for Example 10.4 in WPM Input Format

The following shows the file generated by RWPM-RT for solving Example 10.4 in DIMACS CNF input format.

```
c    #agent: 4 #rule: 5
c    the maximum cost: 7
p    wcnf 9 21 8
c
c    soft clause
1    −1 0
4    2 0
2    3 0
c
c    hard clause generated from dummy rules
8    1 4 5 0
c
c    hard clauses generated from compS
8    −1 −2 6 0
8    −6 1 0
8    −6 2 0
8    −3 −5 7 0
8    −7 3 0
8    −7 5 0
8    −4 −5 8 0
8    −8 4 0
8    −8 5 0
c
c    hard clauses generated from compD
8    −3 −4 −9 0
c
```

Applied Satisfiability: Cryptography, Scheduling, and Coalitional Games, First Edition. Xiaojuan Liao and Miyuki Koshimura.

```
c   hard clauses generated from incomp
8   −1 −3 0
8   −1 −4 0
8   −1 −5 0
8   −2 −3 0
8   −2 −4 0
8   −2 −5 0
c
c   transitive laws
8   −7 −8 9 0
```

Appendix P

Comparative Analysis of RWPM and I-RWPM

The following content in this section comes from our paper [1] and we are including it here to provide a detailed comparison of RWPM and I-RWPM, highlighting their performance differences.

First, we quantitatively analyze the number of clauses for representing transitive laws in RWPM and I-RWPM. Table P.1 lists the notations used in the analysis.

In RWPM, any three rules in $R \cup R_d$ generate three hard clauses for representing transitive laws, thus the total number of hard clauses for transitive laws is

$$T_r = 3 \cdot C^3_{n+n_d} = \left(n + n_d\right) \cdot \left(n + n_d - 1\right) \cdot \left(n + n_d - 2\right)/2 \tag{P.1}$$

where n_d is the cardinality of R_d, and $C^3_{n+n_d}$ is the number of three-combinations of $n + n_d$.

By contrast, in I-RWPM, transitive laws are only created by rules with particular relations over $R_c \cup R_{dc}$. More specifically, as specified in case (4) of Definition 11.2, a hard clause is introduced for r_i and r_j if and only if the following three conditions are all satisfied: (i) r_i and r_j are compatible on different coalitions or independent. (ii) There exists at least one rule compatible with r_i on the same coalition. (iii) There exists a rule r_k compatible with r_j on the same coalition. Therefore, the number of hard clauses for transitive laws in I-RWPM is estimated as follows:

$$\begin{aligned} &T'_r \leq (c + d) \cdot C^2_{n_c+n_{dc}} \cdot \left(n_c + n_{dc} - 2\right) \cdot a \\ &\quad \Longrightarrow T'_r \leq \left(n_c + n_{dc}\right) \cdot \left(n_c + n_{dc} - 1\right) \cdot \left(n_c + n_{dc} - 2\right) \cdot (c + d) \cdot a/2 \end{aligned} \tag{P.2}$$

$$\frac{T'_r}{T_r} \leq \frac{\left(n_c + n_{dc}\right) \cdot \left(n_c + n_{dc} - 1\right) \cdot \left(n_c + n_{dc} - 2\right) \cdot (c + d) \cdot a}{\left(n + n_d\right) \cdot \left(n + n_d - 1\right) \cdot \left(n + n_d - 2\right)}$$

Clearly, $n_c \leq n$ and $n_{dc} \leq n_d$ hold, thus T'_r/T_r is estimated as follows:

$$\frac{T'_r}{T_r} \leq (c + d) \cdot a \tag{P.3}$$

Equation (P.3) suggests that the ratio of T'_r to T_r is no larger than $(c + d) \cdot a$. Since $(c + d) + a \leq 1$, it is evident that $(c + d) \cdot a \leq (1 - a) \cdot a \leq 1/4$. Therefore, the

Applied Satisfiability: Cryptography, Scheduling, and Coalitional Games, First Edition. Xiaojuan Liao and Miyuki Koshimura.

Table P.1 Notations and descriptions used in theoretical comparison

Notation	Description
R	A finite set of rules in an MC-net.
R_c	A set of rules that excludes all freelance rules in R.
R_d	A set of dummy rules created over R.
R_{dc}	A set of dummy rules created over R_c.
n	Total number of rules in R. $n = \|R\|$.
n_c	Number of rules in R_c. $n_c = \|R_c\|$.
n_d	Number of rules in R_d. $n_d = \|R_d\|$.
n_{dc}	Number of rules in R_{dc}. $n_{dc} = \|R_{dc}\|$.
T_r	Number of hard clauses for representing transitive laws in RWPM.
T_r'	Number of hard clauses for representing transitive laws in I-RWPM.
a	The probability that they are compatible on the same coalition.
b	The probability that two rules are incompatible.
c	The probability that two rules are compatible on different coalitions.
d	The probability that two rules are independent.
	Naturally, $a + b + c + d = 1$.

upper bound of T_r'/T_r could be calculated:

$$\frac{T_r'}{T_r} \le \frac{1}{4}$$

This indicates that the number of hard clauses generated for transitive laws in I-RWPM could be reduced by at least a quarter compared with that in RWPM.

Next, we investigate by experiment how a, c, and d change with the increase of *#rules*. In this way, the trend of T_r'/T_r could be captured. For clarity, the rule generation method is summarized as follows: A coalition is initially formed with one randomly selected agent. Subsequently, additional agents are added with a probability of α until no more agents are added or the coalition includes all agents. The value assigned to each rule is randomly selected from a uniform distribution between 0 and the number of agents in the rule. Additionally, for coalitions containing more than one agent, there is a probability p that an agent is converted from positive to negative. Independently, there is a probability q that the coalition's value is converted from positive to negative. The experimental settings follow those in Chapters 8, 9, and 10, i.e., $\alpha = 0.55$, $p = 0.2$, $q = 0.2$, and the number of rules is equal to the number of agents (*#rules* = *#agents*). Results are exhibited in Figure P.1.

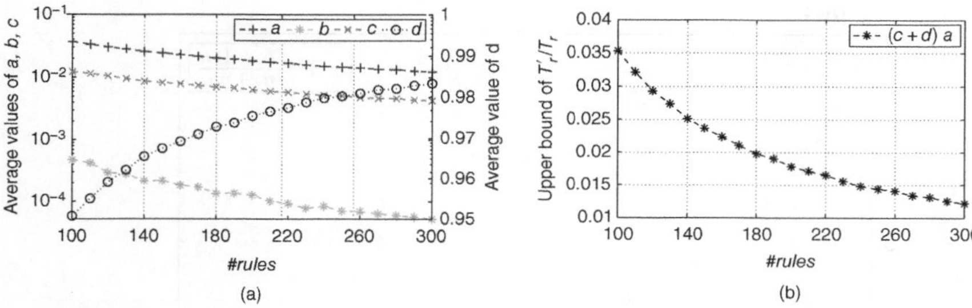

Figure P.1 Values of parameters that affect T_r'/T_r (a) Average values of a, b, c, and d and (b) estimation of the upper bound of T_r'/T_r (Source: Xiaojuan Liao et al. (2019)/with permission of Constraints)

As depicted in Figure P.1a,[1] as *#rules* grows, a and c are decreased while d is increased. In this condition, it is hard to tell whether $(c + d) \cdot a$ gets larger or smaller. Then, we compute $(c + d) \cdot a$ in Figure P.1b, which shows the trend of the upper bound of T_r'/T_r with the increase of *#rules*. It is evident that the value of $(c + d) \cdot a$ monotonously declines as *#rules* goes up. In particular, when *#rules* = 100, the ratio of T_r' to T_r is around 0.035. By contrast, when *#rules* = 300, T_r'/T_r is below 0.015. This suggests that with the increase of *#rules*, the upper bound of T_r'/T_r gets lower and I-RWPM would be more efficient.

Now let us get back to Equations (P.1) and (P.2), which indicate that the magnitudes of T_r' and T_r are $O\left(\left(n_c + n_{dc}\right)^3\right)$ and $O\left(\left(n + n_d\right)^3\right)$, respectively. Apparently, $n_c \leq n$ and $n_{dc} \leq n_d$. Thus the magnitude of T_r' is no larger than that of T_r, although the orders are both 3. In addition, it is noticeable that in Figure P.1b, the ratio of T_r' to T_r is extremely small and such value steadily decreases as the increase of *#rules*. Motivated by the huge numerical difference between T_r' and T_r and the continuous decrease of T_r'/T_r, we conjecture that the actual magnitude of T_r' may be not as great as $O\left(\left(n_c + n_{dc}\right)^3\right)$ due to the influence of the factor $(c + d) \cdot a$ in Eq. (P.2).

In order to verify our guess, we compare the value of $(c + d) \cdot a$ and $\left(n_c + n_{dc}\right)$ based on the current experimental settings and examine whether $(c + d) \cdot a / \left(n_c + n_{dc}\right)$ approaches 0 as *#rules* rises. Results are depicted in Figure P.2.

Evidently, with the increase of *#rules*, the curve of $(c + d) \cdot a / \left(n_c + n_{dc}\right)$ declines steadily. To be specific, when *#rules* = 100, the ratio of $(c + d) \cdot a$ to $\left(n_c + n_{dc}\right)$ is approximately 3×10^{-4}, while when *#rules* = 300, the value falls below 0.5×10^{-4}. The downward trend shown in Figure P.2 is expected to continue for larger *#rules*,

1 In Figure P.1a, the value of d, i.e., the probability that two rules are independent, is quite large. We explain the reason why d is extremely large from a theoretic view in Appendix S.

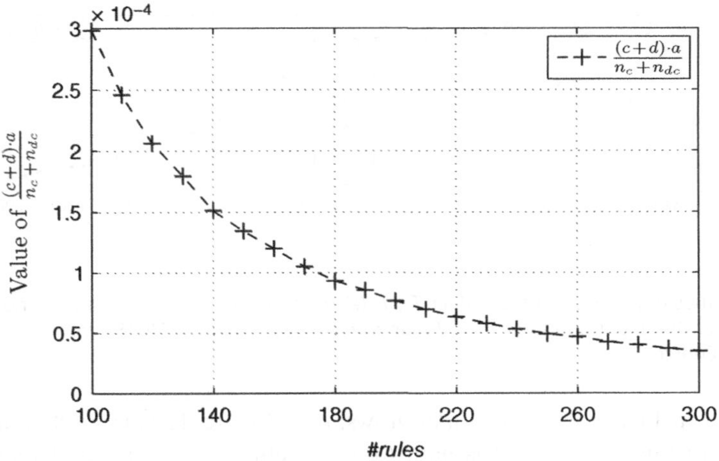

Figure P.2 Trend of $\frac{(c+d)\cdot a}{(n_c+n_{dc})}$ with growing *#rules* (Source: Xiaojuan Liao et al. (2019)/with permission of Constraints)

thus we infer that:

$$lim_{n\to\infty}\frac{(c+d)\cdot a}{n_c+n_{dc}}=0$$

Consequently, based on Eq. (P.2), it can be surmised that with the current parameter settings, T_r' has the order strictly smaller than 3.

In the following, we make a general analysis on the condition which implies that the order of magnitude of T_r' is strictly smaller than 3. Let us assume that the expected value of the number of rules which are compatible with a rule on the same coalition is determined by $\mathbf{x}$ and given by a function $\alpha(\mathbf{x})$ where $\mathbf{x}$ is a tuple of n, the number of positive (or negative) literals in the rule, and so on. Obviously, $\alpha(\mathbf{x}) \leq n_c + n_{dc} - 1$.

The following expression for the probability a holds:

$$\begin{aligned} a &= \frac{\sum_{i=1}^{n_c+n_{dc}} \alpha(\mathbf{x}_i)}{2} \Big/ C^2_{n_c+n_{dc}} \\ &= \frac{\sum_{i=1}^{n_c+n_{dc}} \alpha(\mathbf{x}_i)}{2} \Big/ \frac{(n_c+n_{dc})\cdot(n_c+n_{dc}-1)}{2} \\ &= \frac{\sum_{i=1}^{n_c+n_{dc}} \alpha(\mathbf{x}_i)}{(n_c+n_{dc})\cdot(n_c+n_{dc}-1)} \end{aligned} \tag{P.4}$$

where $\alpha(\mathbf{x}_i)$ is the number of rules compatible on the same coalition with rule r_i. Clearly, $(c+d)/2 < 1$. By substituting Eq. (P.4) into (P.2), we have:

$$T'_r < (n_c + n_{dc}) \cdot (n_c + n_{dc} - 1) \cdot (n_c + n_{dc} - 2) \cdot \frac{\sum_{i=1}^{n_c+n_{dc}} \alpha(\mathbf{x}_i)}{(n_c + n_{dc}) \cdot (n_c + n_{dc} - 1)}$$

$$\Longrightarrow T'_r < (n_c + n_{dc} - 2) \cdot \sum_{i=1}^{n_c+n_{dc}} \alpha(\mathbf{x}_i) \tag{P.5}$$

Thus, the condition that makes the magnitude of T'_r strictly smaller than $O\left((n_c + n_{dc})^3\right)$ is expressed by either of formulas in Eq. (P.6):

$$\lim_{(n_c+n_{dc})\to\infty} \frac{\sum_{i=1}^{n_c+n_{dc}} \alpha(\mathbf{x}_i)}{(n_c + n_{dc})^2} = 0 \quad \text{or} \quad \lim_{(n_c+n_{dc})\to\infty} \frac{\overline{\alpha(\mathbf{x}_i)}}{n_c + n_{dc}} = 0 \tag{P.6}$$

where $\overline{\alpha(\mathbf{x}_i)}$ is the average of $\alpha(\mathbf{x}_i)$. Formally,

$$\overline{\alpha(\mathbf{x}_i)} = \frac{\sum_{i=1}^{n_c+n_{dc}} \alpha(\mathbf{x}_i)}{n_c + n_{dc}}$$

Equation (P.6) indicates that if the probability that a rule is compatible with a specific rule on the same coalition approaches 0 as the total number of rules (including dummy rules for negative value rules) grows, then the order of magnitude of T'_r is strictly smaller than 3.

Reference

1 Xiaojuan Liao, Miyuki Koshimura, Kazuki Nomoto, Suguru Ueda, Yuko Sakurai, and Makoto Yokoo. Improved WPM encoding for coalition structure generation under mc-nets. *Constraints*, 24(1): 25–55, 2019.

Appendix Q

Complete File Generated by I-RWPM for Example 11.3 in WPM Input Format

The following shows the file generated by RWPM-RT for solving Example 11.3 in DIMACS CNF input format.

```
c    #agent: 4 #rule: 4
c    the maximum cost: 8
p    wcnf 7 13 9
c
c    soft clause
2    1 0
2    2 0
1    3 0
3    4 0
c
c    hard clauses generated from compS
9    −1 −2 5 0
9    −5 1 0
9    −5 2 0
9    −1 −3 6 0
9    −6 1 0
9    −6 3 0
c
c    hard clauses generated from compD
9    −2 −3 −7 0
c
c    hard clauses generated from incomp
9    −2 −4 0
c
c    transitive laws
9    −5 −6 7 0
```

Applied Satisfiability: Cryptography, Scheduling, and Coalitional Games, First Edition. Xiaojuan Liao and Miyuki Koshimura.

Appendix R

Complete Files Generated by I-RWPM and RWPM-RT for Example 11.4 in WPM Input Format

The following shows the file generated by I-RWPM for solving Example 11.4 in DIMACS CNF input format.

```
c    #agent: 4 #rule: 4
c    the maximum cost: 8
p    wcnf 7 12 9
c
c    soft clause
3    1 0
2    2 0
1    3 0
2    4 0
c
c    hard clauses generated from compS
9    −2 −4 5 0
9    −5 2 0
9    −5 4 0
9    −3 −4 6 0
9    −6 3 0
9    −6 4 0
c
c    hard clauses generated from compD
c
c    hard clauses generated from incomp
9    −1 −2 0
c
c    transitive laws
9    −5 −6 7 0
```

Applied Satisfiability: Cryptography, Scheduling, and Coalitional Games, First Edition. Xiaojuan Liao and Miyuki Koshimura.

The following shows the file generated by RWPM-RT for solving Example 11.4 in DIMACS CNF input format.

```
c   #agent: 4 #rule: 4
c   the maximum cost: 8
p   wcnf 10 17 9
c
c   soft clause
3   1 0
2   2 0
1   3 0
2   4 0
c
c   hard clauses generated from compS
9   −2 −4 5 0
9   −5 2 0
9   −5 4 0
9   −3 −4 6 0
9   −6 3 0
9   −6 4 0
c
c   hard clauses generated from compD
9   −1 −3 −7 0
9   −1 −4 −8 0
c
c   hard clauses generated from incomp
9   −1 −2 0
c
c   transitive laws
9   −8 −6 7 0
9   −9 −5 8 0
9   −7 −6 8 0
9   −5 −6 10 0
```

Appendix S

Theoretical Analysis on *d*

This part explains the reason why d is quite large in our experiments by theoretical derivation, described as follows.

Let $m \in \mathbb{Z}^+$ and $\overline{m} \in \mathbb{Z}^+$ be the total number of agents in an MC-net and the average number of agents in a rule, respectively. If $\overline{m} \leq m/2$, we have

$$d = \frac{C_m^{\overline{m}} C_{m-\overline{m}}^{\overline{m}}}{C_m^{\overline{m}} C_m^{\overline{m}}} = \frac{C_{m-\overline{m}}^{\overline{m}}}{C_m^{\overline{m}}}$$

where $C_m^{\overline{m}}$ is the number of $\overline{m}$-combinations of m.

Next, we compute $\overline{m}$. Let random variable X represent the number of agents contained in a rule, $X = \{1, 2, \ldots, m\}$, where m is the number of agents in an MC-net. According to the method of generating instances illustrated in Section 11.4, each condition is created with one random agent, then a new random agent is repeatedly added with the probability of α until an agent is not added or the rule includes all agents. Based on this method, the probability that $X = t$ is represented as follows:

$$P(X) = \begin{cases} \alpha^{t-1}(1-\alpha) & \text{if } t < m, \\ \alpha^{m-1} & \text{otherwise} \end{cases}$$

Then the expectation of X is $E(X) = \sum_{t=1}^{m-1} t\alpha^{t-1}(1-\alpha) + m\alpha^{m-1}$. By simplifying the formula, we have

$$E(X) = \frac{1-\alpha^m}{1-\alpha}$$

Applied Satisfiability: Cryptography, Scheduling, and Coalitional Games, First Edition. Xiaojuan Liao and Miyuki Koshimura.

Thus, $\overline{m}$ can be estimated by the expected value of X, i.e., $\overline{m} = \frac{1-\alpha^m}{1-\alpha}$.[1] In our experiments, $\alpha = 0.55$, and m ranges from 100 to 300. With this experimental setting, it is immediate that $d \approx 0.96$. If a smaller d is required, then we can make the value of α larger. Specifically, when α reaches 0.75, $\overline{m} \approx 4$ and $d \approx 0.85$.

1 To be more precise, depending on the number of agents and the value of α, the average number $\overline{m}$ will either be $\lfloor \frac{1-\alpha^m}{1-\alpha} \rfloor$ or $\lceil \frac{1-\alpha^m}{1-\alpha} \rceil$. However, to ease the discussion without loss of generality, we will assume that $\overline{m}$ is simply $\frac{1-\alpha^m}{1-\alpha}$.

Index

Applied Satisfiability: Cryptography, Scheduling, and Coalitional Games, First Edition. Xiaojuan Liao and Miyuki Koshimura.

Printed and bound by CPI Group (UK) Ltd, Croydon, CR0 4YY
17/09/2026
14964329-0002